TRANSPORTATION ISSUES, POLICIES AND R&D

OIL AND GAS TRANSPORTATION

PIPELINE AND RAIL INFRASTRUCTURE ISSUES

TRANSPORTATION ISSUES, POLICIES AND R&D

Additional books in this series can be found on Nova's website under the Series tab.

Additional e-books in this series can be found on Nova's website under the e-book tab.

TRANSPORTATION ISSUES, POLICIES AND R&D

OIL AND GAS TRANSPORTATION: PIPELINE AND RAIL INFRASTRUCTURE ISSUES

ELTON SIMMONS
EDITOR

nova publishers
New York

Library of Congress Cataloging-in-Publication Data

ISBN: 978-1-63463-841-8

Published by Nova Science Publishers, Inc. † New York

CONTENTS

PREFACE

Increased oil and gas production presents challenges for transportation infrastructure because some of this increase is in areas with limited transportation linkages. Technology advancements such as horizontal drilling and hydraulic fracturing (pumping water, sand, and chemicals into wells to fracture underground rock formations and allow oil or gas to flow) have allowed companies to extract oil and gas from shale and other tight geological formations. This book examines overall challenges that increased oil and gas production may pose for transportation infrastructure; specific pipeline safety risks and how the U.S. Department of Transportation (DOT) is addressing them; and specific rail safety risks and how DOT is addressing them.

Chapter 1 - Technology advancements such as horizontal drilling and hydraulic fracturing (pumping water, sand, and chemicals into wells to fracture underground rock formations and allow oil or gas to flow) have allowed companies to extract oil and gas from shale and other tight geological formations. As a result, oil and gas production has increased more than fivefold from 2007 through 2012. DOT oversees the safety of the U.S. transportation system.

GAO was asked to review oil and gas transportation infrastructure issues. This report examines (1) overall challenges that increased oil and gas production may pose for transportation infrastructure, (2) specific pipeline safety risks and how DOT is addressing them, and (3) specific rail safety risks and how DOT is addressing them. GAO analyzed federal transportation infrastructure and safety data generally from 2008 to 2012 or 2013 (as available), reviewed documents, and interviewed agency, industry, and safety stakeholders, as well as state and industry officials in states with large-scale shale oil and gas development.

Chapter 2 - North America is experiencing a boom in crude oil supply, primarily due to growing production in the Canadian oil sands and the recent expansion of shale oil production from the Bakken fields in North Dakota and Montana as well as the Eagle Ford and Permian Basins in Texas. Taken together, these new supplies are fundamentally changing the U.S. oil supply-demand balance. The United States now meets 66% of its crude oil demand from production in North America, displacing imports from overseas and positioning the United States to have excess oil and refined products supplies in some regions.

The rapid expansion of North American oil production has led to significant challenges in transporting crudes efficiently and safely to domestic markets—principally refineries—using the nation's legacy pipeline infrastructure. In the face of continued uncertainty about the prospects for additional pipeline capacity, and as a quicker, more flexible alternative to new pipeline projects, North American crude oil producers are increasingly turning to rail as a means of transporting crude supplies to U.S. markets. According to rail industry officials, U.S. freight railroads are estimated to have carried 434,000 carloads of crude oil in 2013 (roughly equivalent to 300 million barrels), compared to 9,500 carloads in 2008. In 2014, 650,000 carloads of crude oil are expected to be carried. Crude imports by rail from Canada have increased more than 20-fold since 2011. The amount of oil transported by rail may also be influenced by a tight market for U.S.-built tankers.

While oil by rail has demonstrated benefits with respect to the efficient movement of oil from producing regions to market hubs, it has also raised significant concerns about transportation safety and potential impacts to the environment. The most recent data available indicate that railroads consistently spill less crude oil per ton-mile transported than other modes of land transportation. Nonetheless, safety and environmental concerns have been underscored by a series of major accidents across North America involving crude oil transportation by rail—including a catastrophic fire that caused numerous fatalities and destroyed much of Lac Mégantic, Quebec, in 2013. Following that event, the U.S. Department of Transportation issued a safety alert warning that the type of crude oil being transported from the Bakken region may be more flammable than traditional heavy crude oil.

Legislation introduced in Congress following the Lac Mégantic disaster would require railroads to have at least two crew members aboard all trains. In addition, policy makers are discussing regulatory changes involving tank car design, prevention of derailments, and selection of preferred routes for transporting oil by rail. Congress may evaluate these changes in the reauthorization of the Rail Safety Improvement Act of 2008 (P.L. 110-432).

Chapter 3 - Nearly half a million miles of pipeline transporting natural gas, oil, and other hazardous liquids crisscross the United States. While an efficient and fundamentally safe means of transport, many pipelines carry materials with the potential to cause public injury and environmental damage. The nation's pipeline networks are also widespread and vulnerable to accidents and terrorist attack. Recent pipeline accidents in Marshall, MI, San Bruno, CA, Allentown, PA, and Laurel, MT, have heightened congressional concern about pipeline risks and drawn criticism from the National Transportation Safety Board (NTSB). Both government and industry have taken numerous steps to improve pipeline safety and security over the last 10 years. Nonetheless, while many stakeholders agree that federal pipeline safety programs have been on the right track, the spate of recent pipeline incidents suggest there continues to be significant room for improvement. Likewise, the threat of terrorist attacks, especially cyberattacks on pipeline control systems, remains a concern.

The federal pipeline safety program is authorized through the fiscal year ending September 30, 2015, under the Pipeline Safety, Regulatory Certainty, and Job Creation Act of 2011 (P.L. 112- 90), which was signed by President Obama on January 3, 2012. The act contains a broad range of provisions addressing pipeline safety and security. Among the most significant are provisions that could increase the number of federal pipeline safety inspectors, require automatic shutoff valves for transmission pipelines, mandate verification of maximum allowable operating pressure for gas transmission pipelines, increase civil penalties for pipeline safety violations, and mandate reviews of diluted bitumen pipeline regulation.

Both government and industry have taken numerous steps to improve pipeline safety and security over the last 10 years. Nonetheless, the NTSB has identified improvement of federal pipeline safety oversight as a "top ten" priority for 2013. The leading pipeline industry associations have concurred. Whether renewed efforts by industry, combined with additional oversight by federal agencies, will further enhance the safety and security of U.S. pipelines remains to be seen.

As Congress oversees the federal pipeline safety program and the federal role in pipeline security, key issues of focus may be pipeline agency staff resources, automatic pipeline shutoff valves, penalties for safety violations, safety regulations for oil sands crudes, and the possible need for pipeline security regulations, among other concerns. In addition to these specific issues,

Congress may assess how the various elements of U.S. pipeline safety and security activity fit together in the nation's overall strategy to protect transportation infrastructure. Pipeline safety and security necessarily involve many groups: federal agencies, oil and gas pipeline associations, large and small pipeline operators, and local communities. Reviewing how these groups work together to achieve common goals could be an oversight challenge for Congress.

In: Oil and Gas Transportation
Editor: Elton Simmons
ISBN: 978-1-63463-841-8

Chapter 1

OIL AND GAS TRANSPORTATION: DEPARTMENT OF TRANSPORTATION IS TAKING ACTIONS TO ADDRESS RAIL SAFETY, BUT ADDITIONAL ACTIONS ARE NEEDED TO IMPROVE PIPELINE SAFETY*

United States Government Accountability Office

WHY GAO DID THIS STUDY

Technology advancements such as horizontal drilling and hydraulic fracturing (pumping water, sand, and chemicals into wells to fracture underground rock formations and allow oil or gas to flow) have allowed companies to extract oil and gas from shale and other tight geological formations. As a result, oil and gas production has increased more than fivefold from 2007 through 2012. DOT oversees the safety of the U.S. transportation system.

GAO was asked to review oil and gas transportation infrastructure issues. This report examines (1) overall challenges that increased oil and gas production may pose for transportation infrastructure, (2) specific pipeline safety risks and how DOT is addressing them, and (3) specific rail safety risks

* This is an edited, reformatted and augmented version of a United States Government Accountability Office publication, No. GAO-14-667, dated August 2014.

and how DOT is addressing them. GAO analyzed federal transportation infrastructure and safety data generally from 2008 to 2012 or 2013 (as available), reviewed documents, and interviewed agency, industry, and safety stakeholders, as well as state and industry officials in states with large-scale shale oil and gas development.

What GAO Recommends

DOT should move forward with a proposed rulemaking to address safety risks—including emergency response planning—from newer gathering pipelines. DOT generally concurred with the recommendation and stated that it is developing a rulemaking to revise its pipeline safety regulations.

What GAO Found

Increased oil and gas production presents challenges for transportation infrastructure because some of this increase is in areas with limited transportation linkages. For example, insufficient pipeline capacity to transport crude oil has resulted in the increased use of rail, truck, and barge to move oil to refineries, according to government and industry studies and publications GAO reviewed. These transportation limitations and related effects could pose environmental risks and have economic implications. For instance, natural gas produced as a byproduct of oil is burned—a process called flaring—by operators due, in part, to insufficient pipelines in production areas. In a 2012 report, GAO found that flaring poses a risk to air quality as it emits carbon dioxide, a greenhouse gas linked to climate change, and other air pollutants. In addition, flaring results in the loss of a valuable resource and royalty revenue.

Due to the increased oil and gas production, construction of larger, higher-pressure gathering pipelines (pipelines that transport products to processing facilities and other long-distance pipelines) has increased. However, these pipelines, if located in rural areas, are generally not subject to U.S. Department of Transportation (DOT) safety regulations that apply to other pipelines, including emergency response requirements. Historically, gathering pipelines were smaller and operated at lower pressure and thus posed less risk than long-distance pipelines. But the recent increase in their size and pressure raises safety concerns because they could affect a greater area in the event of an incident. In 2011, DOT began a regulatory proceeding to address the safety risks of gathering pipelines, but it has not proposed new regulations. Although states may regulate gathering pipelines, an association of state pipeline regulators' report on state pipeline oversight shows that most states do not currently regulate gathering pipelines in rural areas.

Crude oil carloads moved by rail in 2012 increased by 24 times over that moved in 2008. Such an increase raises specific concerns about testing and packaging of crude oil, use of unit trains (trains of about 80 to 120 crude oil cars), and emergency response preparedness. Crude oil shippers are required to identify their product's hazardous properties, including flammability, before packaging the oil in an authorized tank car. DOT has issued safety alerts on the importance of proper testing and packaging of crude oil. However, industry stakeholders said that DOT's guidance on this issue is vague and that clarity about the type and frequency of testing is needed. In July 2014, DOT proposed new regulations for crude oil shippers to develop a product-testing program subject to DOT's review. Additionally, unit trains, which can carry 3 million or more gallons of crude oil and travel to various locations through the country, are not covered under DOT's comprehensive emergency response planning requirements for transporting crude oil by rail because the requirements currently only apply to individual tank cars and not unit trains. This raises concerns about the adequacy of emergency response preparedness, especially in rural areas where there may be fewer resources to respond to a serious incident. Also in July 2014, DOT sought public comment on potential options for addressing this gap in emergency response planning requirements for transporting crude oil by rail.

ABBREVIATIONS

AAR	Association of American Railroads
DOT	Department of Transportation
EIA	Energy Information Administration
FRA	Federal Railroad Administration
INGAA	Interstate Natural Gas Association of America
LNG	liquefied natural gas
Mcf	thousand cubic feet
MMBtu	million British thermal units
NOx	nitrogen oxides
NTSB	National Transportation Safety Board

PHMSA	Pipeline and Hazardous Materials Safety Administration
PSI	pounds per square inch
STB	Surface Transportation Board
Tcf	trillion cubic feet

* * *

August 21, 2014

The Honorable John D. Rockefeller, IV
Chairman
The Honorable John Thune
Ranking Member
Committee on Commerce, Science, and Transportation
United States Senate

The Honorable Ron Wyden
United States Senate

U.S. production of oil and gas resources has increased in recent years, driven in part by improvements in technologies. Oil and gas resources contained in underground shale formations were previously considered to be inaccessible because traditional techniques did not yield sufficient amounts for economically viable production. The application of horizontal drilling techniques and hydraulic fracturing—a process that injects a combination of water, sand, and chemical additives under high pressure to create and maintain fractures in underground rock formations that allow oil and natural gas to flow—have increased U.S. crude oil and natural gas production dramatically.

This rapid expansion of domestic oil and gas production has also changed dynamics for transporting these products to the market and has raised questions about safety. While pipelines transport the majority of oil and gas in the United States, recent development of crude oil in parts of the country underserved by pipeline has led shippers to use other modes, with rail seeing the largest percentage increase. Although pipeline operators and railroads have generally good safety records, the increased transportation of these flammable hazardous materials creates the potential for serious incidents. The explosion and fire caused by the July 2013 derailment of a crude oil train in Lac-Mégantic, Quebec killed 47 people and extensively damaged the city's downtown area, highlighting the consequences that may result from such incidents. The U.S. Department of Transportation (DOT) is responsible for ensuring the safety of the U.S. transportation system, including protecting people and the environment from the risks of transporting hazardous materials by pipeline, rail, and other modes. In particular, DOT's Pipeline and Hazardous Materials Safety Administration (PHMSA) has responsibility for pipeline safety oversight as well as hazardous materials transportation safety oversight for other transportation modes, including rail.

You requested that we examine the impact of shale oil and gas development on transportation infrastructure and safety. We are providing a broad overview of transportation infrastructure impacts and a closer look at the infrastructure changes and associated safety issues with pipeline and rail.[1] Specifically, this report addresses:

1) challenges, if any, that increased domestic oil and gas production poses for U.S. transportation infrastructure and examples of associated risks and implications;
2) how pipeline infrastructure has changed as a result of increased oil and gas production, the key related safety risks, and to what extent DOT has addressed these risks; and
3) how rail infrastructure has changed as a result of increased oil production, the key related safety risks, and to what extent DOT has addressed these risks.

To identify challenges increased domestic oil and gas production poses for U.S. transportation infrastructure and the associated implications, we reviewed and synthesized information from studies and other publications from federal, state, and tribal government agencies; industry; academics; and other organizations. We identified these studies and publications by conducting a search of web-based databases and other resources containing general academic articles, government resources, and "gray literature."[2] We believe the studies and publications identified through our literature search provide key examples of transportation infrastructure limitations and associated implications. In addition, we analyzed data from the U.S. Department of Energy's Energy Information Administration (EIA) on oil and gas produced from 2007 to 2012. To assess the reliability of these data, we examined EIA's published methodology for collecting this information and found the data sufficiently reliable for the purposes of this report.

To determine how pipeline infrastructure has changed as a result of increased oil and gas production, we analyzed PHMSA data on pipeline construction from January 1, 2010 through December 31, 2012 and interviewed DOT officials and industry representatives, including pipeline operators. To determine how rail infrastructure has changed, we analyzed the Surface Transportation Board's (STB) data for calendar years 2008 through 2012 on crude oil shipments by rail and interviewed DOT officials and industry representatives, including railroads. To identify the key safety risks related to changes in pipeline and rail infrastructure, we analyzed PHMSA data from January 1, 2008 through December 31, 2013 on pipeline and rail incidents, reviewed documents submitted as part of DOT's rulemaking on rail safety, and interviewed DOT officials and representatives from safety organizations, emergency responder associations, and industry. We assessed the reliability of PHMSA's data on pipeline construction and pipeline and rail incidents and STB's data on crude oil shipments by rail by reviewing documentation about the data sources, interviewing agency officials about how the data were collected, and reviewing related internal controls. We also reviewed some of the data for potential inconsistencies through testing and comparing the data to publicly available sources of similar information. We concluded that these data were sufficiently reliable for the purposes used in our report. Additionally, to examine infrastructure impacts and safety risks closely associated with transporting shale oil and gas, we interviewed officials and reviewed related documents from state oil and gas safety regulatory agencies, transportation departments, industry associations and oil and gas transportation companies (such as pipeline operators, railroads, and operators of rail loading terminals) in four states: North Dakota, Pennsylvania, Texas, and West Virginia. We selected these states because of their significant shale oil and gas development and varying geographic locations. To evaluate to what extent DOT has addressed safety risks, we reviewed federal laws and regulations and DOT emergency orders and guidance, interviewed DOT officials, and compared DOT's actions against risk-based management principles. See appendix I for a more detailed description of our objectives, scope, and methodology.

We conducted this performance audit from August 2013 to August 2014 in accordance with generally accepted government auditing standards. Those standards require that we plan and perform the audit to obtain sufficient, appropriate evidence to provide a reasonable basis for our findings and conclusions based on our audit objectives. We believe that the evidence obtained provides a reasonable basis for our findings and conclusions based on our audit objectives.

BACKGROUND

Location of Oil and Gas Development in the United States

Oil and natural gas are found in a variety of geologic formations distributed across the country, such as shale and tight sandstone.[3] Shale plays—sets of discovered or undiscovered oil and natural gas accumulations or prospects that exhibit similar geological characteristics—are located within basins, which are large-scale geological depressions, often hundreds of miles across, that also may contain other oil and gas resources. Figure 1 shows the location of shale plays and basins in the contiguous 48 states.

Shale plays can contain oil, natural gas, or both. In addition, a shale gas play may contain “dry” or “wet” natural gas. Dry natural gas is a mixture of hydrocarbon compounds that exists as a gas both underground in the reservoir and during production under standard temperature and pressure conditions. Wet natural gas contains natural gas liquids, or the portion of the hydrocarbon resource that exists as a gas when in natural underground reservoir conditions but that is liquid at surface conditions. The natural gas liquids are typically propane, butane, and ethane and are separated from the produced natural gas at the surface.

Operators may then sell the natural gas liquids, which may give wet shale gas plays an economic advantage over dry gas plays. According to a 2014 EIA publication, operators moved away from the development of shale plays that are primarily dry gas in favor of developing plays with higher concentrations of crude oil and natural gas liquids such as the Eagle Ford in Texas, because given natural gas prices at that time, crude oil and natural gas liquids were more valuable products.[4]

Another advantage of liquid petroleum and natural gas liquids is that they can be transported more easily through different modes of transportation than dry natural gas, which is transported almost entirely by pipelines to markets and consumers.

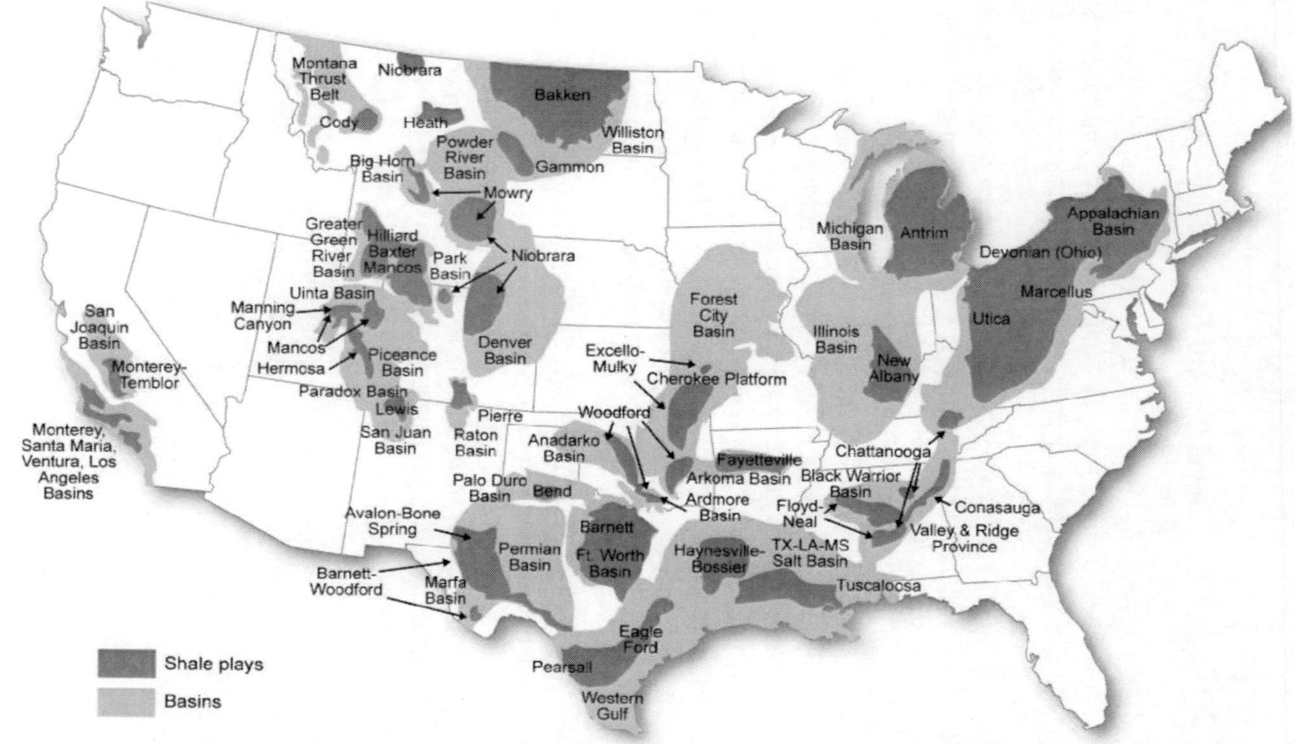

Sources: Energy Information Administration (shale location data); (map) copyright © Corel Crop., all rights reserved. | GAO-14-667.
Note: Tight sandstone basins are found in some of the same basins as shale plays.

Figure 1. Shale Plays and Basins in the Contiguous 48 States.

Oil and Gas Production

In recent years, domestic onshore production of oil and gas has been steadily rising. For example, from 2007 through 2012, annual production from shale and tight sandstone formations increased more than sixfold for oil and approximately fivefold for gas (see fig. 2).

Horizontal drilling and hydraulic fracturing have advanced significantly over the last decade and are largely credited with spurring the boom in oil and gas production in the United States.

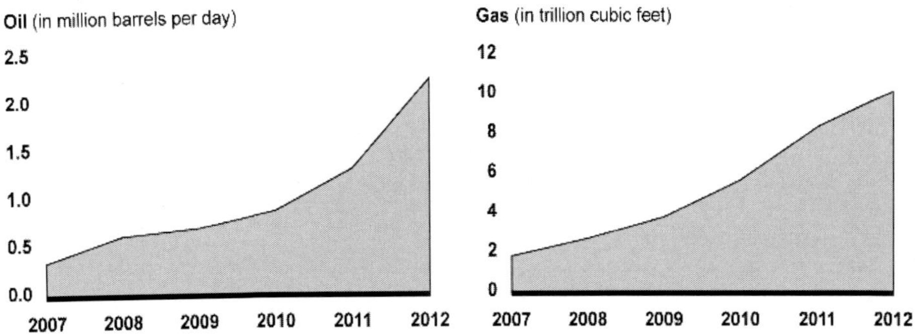

Source: Energy Information Administration. | GAO-14-667.

Figure 2. Increased Domestic Oil and Gas Production from Shale and Tight Sandstone Formations from 2007 to 2012.

Oil: Average domestic crude oil production from shale and tight sandstone formations in 2012 has increased more than sixfold compared with average production in 2007, from 0.34 million barrels per day in 2007 to 2.25 million barrels per day in 2012.

To put this into context, according to EIA data, the United States consumed an average of more than 18 million barrels of petroleum products per day in 2012.[5] According to EIA officials, oil from shale and tight sandstone formations accounts for 31 percent of total U.S. production. According to EIA, increased production in 2012 and 2013 was the largest annual increase since the beginning of U.S. commercial crude oil production in 1859. Much of the increase in crude oil production has been from shale formations, such as the Bakken in North Dakota, the Eagle Ford in Texas, and the Niobrara in Colorado.

According to EIA officials, U.S. production of crude oil is expected to continue to increase—by 48 percent from 2012 to 2019—and will remain above the 2012 level through 2040.

Natural Gas: Domestic natural gas production in 2012 has increased about fivefold compared with production in 2007, from less than 2 trillion cubic feet in 2007 to more than 10 trillion cubic feet in 2012.[6] To put this into context, annual domestic consumption of natural gas was just over 25 trillion cubic feet in 2012, according to EIA data. In September 2012, we found that, assuming current consumption levels without consideration of a specific market price for future gas supplies, the amount of domestic technically recoverable shale gas could provide enough natural gas to supply the nation for the next 14 to 100 years.[7]

Much of the increase in natural gas has been from shale formations, such as the Barnett, Fayetteville, Haynesville, and Marcellus formations.

Transportation Modes

Multiple modes of transportation, including pipeline, rail, highways, and waterways, connect oil and gas production infrastructure (such as wells and processing plants) in shale areas to customers, which include refineries, industrial users, and individual consumers. Additionally, when products switch modes of transportation, oil-loading terminals, sometimes referred to as "transload" terminals, transfer the product from one mode to another, such as when crude oil is transferred from a truck or gathering pipeline to a train. Responsibility for maintaining these modes vary: pipelines and rail are generally privately owned, while highways and waterways are generally public.

Figure 3 illustrates how various transportation modes work together to bring oil and gas from production areas to users.

Approximately 2.5 million miles of pipelines transport roughly two-thirds of domestic energy supplies throughout the United States. These pipelines carry natural gas and hazardous liquids, including crude oil and natural gas liquids from production areas to end users, such as residences and businesses. Gathering pipelines collect produced oil and gas from their source and transport these products to processing facilities and transmission pipelines. Transmission pipelines then transport these products longer distances to users such as residences and businesses.[8]

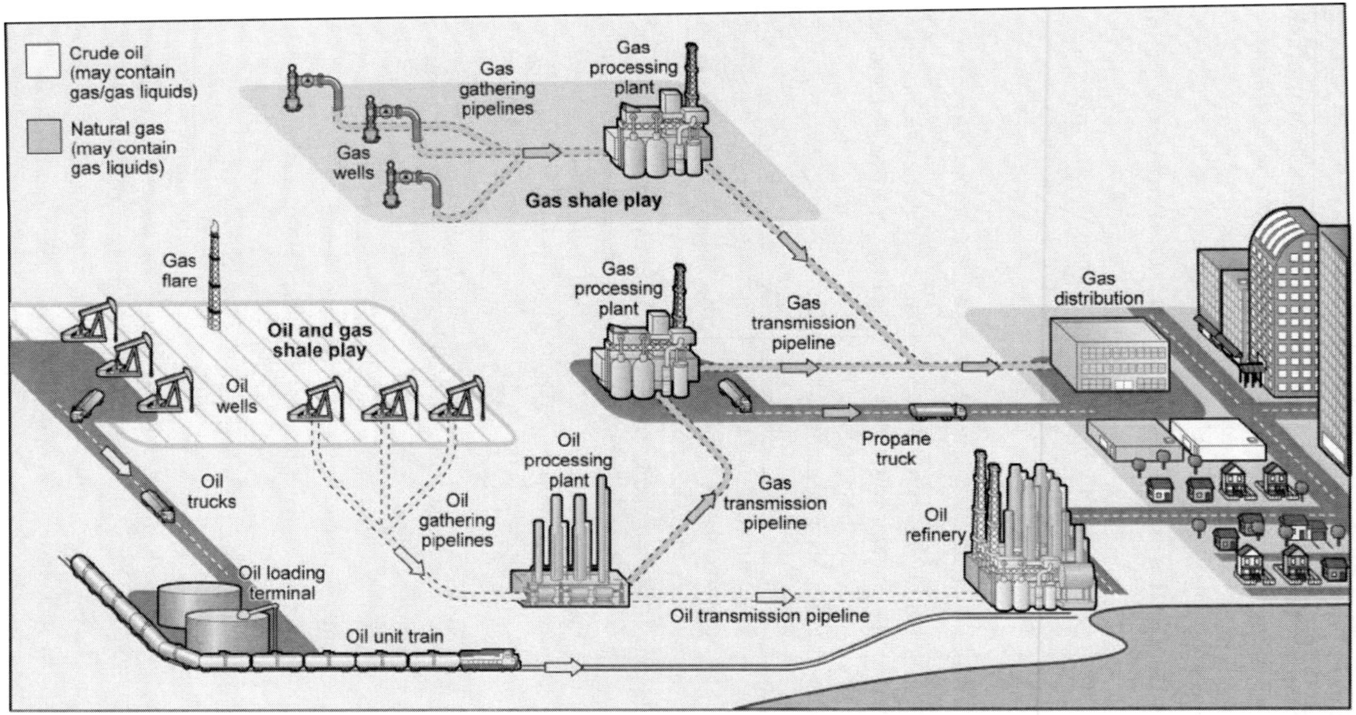

Source: GAO. | GAO-14-667.

Note: This figure illustrates concepts presented in this report. It does not include all modes of transportation for oil and gas products nor does it depict all types of potential end users.

Figure 3. Oil and Gas Transportation from Production to Users.

Distribution pipelines transport natural gas to consumers for use and are not within the scope of this report. Characteristics of gathering and transmission pipelines are described in table 1.

Table 1. Characteristics of U.S. Gathering and Transmission Pipelines

	Gathering pipelines	Transmission pipelines
Estimated number of miles	At least 230,000	More than 400,000
Function	Collect gas or hazardous liquids from production areas and then typically transport these products to processing facilities where they are refined.	Carry gas or hazardous liquid to communities and large-volume users (e.g., factories).
Description	Traditionally, range in diameter from 2 to 12 inches and operate at pressures ranging from 5 to 800 pounds per square inch (psi).	Traditionally over hundreds of miles long, range indiameter from 12 to 42 inches, and operate at pressures ranging from 400 to 1440 psi.

Source: GAO analysis of information from Pipeline and Hazardous Materials Safety Administration. | GAO-14-667.

The U.S. rail network consists of about 200,000 miles of track, which runs mostly through rural areas. The railroad industry is dominated by the seven largest railroads, known as Class I railroads, which collectively accounted for more than 90 percent of annual railroad-freight revenues in 2012. Smaller regional and short-line railroads transport freight shorter distances and can help connect customers in areas not served by the larger railroads. The railroads' national association, the Association of American Railroads (AAR), represents the interests of the industry and works with railroads and other stakeholders to develop industry standards. Crude oil travels by rail in tank cars, commonly DOT-111 tank cars, which are generally owned by shippers or third parties. The DOT-111 is a DOT-specification tank car, meaning it must be built to conform to standards specified in DOT regulation.[9] It is a non-pressurized car that is used to transport a variety of liquid products, including hazardous, flammable materials like crude oil. Terminals, referred to as transload facilities, transfer crude oil from other transportation modes (typically trucks or gathering pipelines) to tank cars for transport by train.

In addition to pipeline and rail, other modes, including barge and truck may transport oil and gas products. For example, barges may transport oil over longer distances on major waterways, such as the Mississippi River, while trucks typically transport oil over short distances to transload facilities. While this report provides a closer look at transportation infrastructure and safety impacts of shale oil and natural gas development on pipeline and rail nationwide, we also discuss highway infrastructure and safety impacts in the four selected states we examined (see app. II for a summary of highway-related impacts).

Department of Transportation

DOT is responsible for ensuring the safe transportation of people and goods through regulations, oversight, inspections, and other efforts, sometimes in partnership with states. Within DOT, PHMSA's Office of Pipeline Safety oversees the safety of pipelines through regulation and an inspection program, which includes over 100 PHMSA inspectors, and also collects information about the location of pipelines. PHMSA also has arrangements with states, which collectively have over 300 inspectors, to assist with overseeing interstate pipelines, intrastate pipelines, or both.[10] PHMSA's current pipeline regulations cover all hazardous liquid (including crude oil) and natural gas transmission pipelines. In addition to minimum safety standards that all transmission pipeline operators must meet, PHMSA employs a risk-based approach to transmission pipeline regulation and requires operators to systematically identify and mitigate risks in "high-consequence areas," which include populated and environmentally sensitive areas.[11] PHMSA also applies this risk-based approach to gathering pipelines and regulates gathering pipelines in non-rural areas, resulting in regulation of approximately 10 percent of the nation's gathering pipelines.[12] Generally, PHMSA retains full responsibility for inspecting interstate pipelines for compliance with its regulations and taking enforcement actions when needed. However, states may be authorized to conduct inspections of interstate pipelines, as well as inspections and associated enforcement for intrastate pipelines. States can also promulgate regulations for intrastate pipelines, including gathering pipelines, even if these pipelines are not covered by PHMSA's federal safety requirements.[13]

PHMSA, through its Office of Hazardous Materials Safety, also regulates shippers and railroads transporting hazardous materials like crude oil by rail and other modes.[14] A memorandum of agreement details how PHMSA works with the other DOT modal agencies to address hazardous-material transportation safety. DOT's other modal administrations have responsibility for safety of their respective modes, such as the Federal Railroad Administration (FRA), which oversees rail safety. FRA enforces its own and PHMSA's safety regulations through inspections by FRA officials and state partners in some states.[15] PHMSA also has hazardous materials inspectors who enforce requirements for hazardous material packaging for transportation. Additionally, PHMSA's regulations include emergency response planning requirements for pipelines and the transportation of crude oil by rail. Specifically, regulations require operators of transmission pipelines and urban gathering pipelines to prepare emergency response plans and coordinate them with emergency responders.[16] Railroads that transport crude oil in tanks larger than 42,000 gallons are required to develop comprehensive oil-spill response plans with additional requirements for contingency planning, ensuring response resources by contract or other means, and training. Railroads are required to submit comprehensive plans to FRA for review.

Otherwise, railroads are required to develop basic response plans, for which there are fewer requirements.[17] Because PHMSA applies a risk-based approach to its transportation oversight, we believe it is appropriate to apply principles of risk-based management to assessing the agency's efforts in this area. Risk-based management has several key characteristics that help to ensure safety, including (1) using information to identify and assess risks; (2) prioritizing risks so that resources may be allocated to address higher risks first; and (3) promoting the use of regulations, policies, and procedures to provide consistency in decision making.[18]

Increased Oil and Gas Production Presents Challenges for Transportation Infrastructure That Could Pose Environmental and Safety Risks and Have Economic Implications

Increased oil and gas production presents challenges for transportation infrastructure because some of the growth in production has been in areas with limited transportation linkages to processing facilities. According to studies and publications we reviewed, infrastructure limitations and related effects could pose environmental and safety risks and have economic implications, including lost revenue and hindered oil and gas production.

Increased Domestic Oil and Gas Production Presents Challenges for Transportation Infrastructure

Though capital investments in U.S. infrastructure for oil and gas transportation, processing, and storage have increased significantly in recent years—by 60 percent from 2008 to 2012, according to a December 2013 industry report—expansions in infrastructure have not kept pace with increased domestic oil and gas production.[19] In the United States, most oil and nearly all natural gas are transported by pipeline.[20] According to EIA data, U.S. refinery receipts of domestic crude oil by pipeline increased almost 25 percent from 2008 to 2012, from 1.6 billion barrels to nearly 2 billion barrels.[21] However, according to a number of studies and publications we reviewed, including a 2013 report from the Fraser Institute, oil and natural gas production in the United States is outpacing the capacity to transport the resources through existing pipeline infrastructure.[22] In February 2013, EIA reported that pipeline capacity to deliver crude oil to a key hub increased by about 815,000 barrels per day from 2010 through 2013; however, the increase has been inadequate to transport crude oil from production sites to refineries. In March 2014, we found that most of the system of crude oil pipelines in the United States was constructed in the 1950s, 1960s, and 1970s to accommodate the needs of the refining sector and demand centers at that time. We also found that, according to Department of Energy officials, this infrastructure was designed primarily to move crude oil from the South to the North, but emerging crude oil production centers in Western Canada, Texas, and North Dakota have strained the existing pipeline infrastructure.[23] For example, according to a 2013 industry publication, oil production exceeded pipeline capacity in North Dakota by about 300,000 barrels of oil per day in the state.[24]

The limited pipeline capacity to transport crude oil has resulted in the increased use of other transportation options, in particular rail, truck, and barge (see fig. 4).

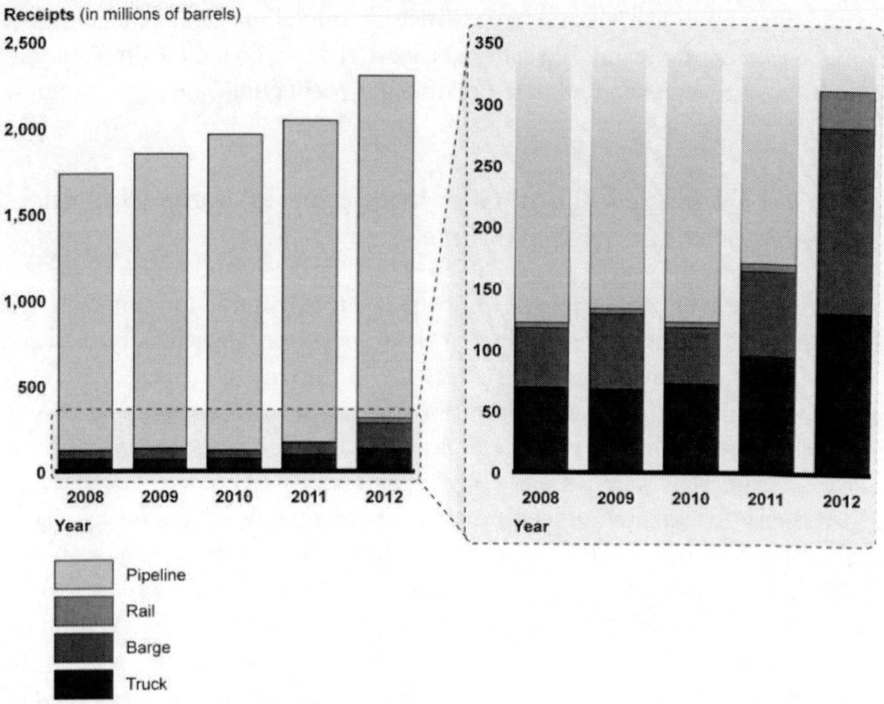

Source: GAO analysis of EIA data. | GAO-14-667.

Note: Oil shipments may involve multiple modes. This figure indicates only the mode used for the last leg of the shipment.

Figure 4. Crude Oil Refinery Receipts by Pipeline, Rail, Truck, and Barge from 2008 to 2012 (Millions of Barrels).

- ***Rail:*** According to a 2014 EIA report, U.S. refinery receipts of domestic crude oil by rail increased more than sevenfold from 2008 to 2012, from 4 million barrels to 30 million barrels.[25] The increased use of rail for transporting crude oil is due to the increases in crude oil production in North Dakota, Texas, and other states, which have exceeded the capacity of existing pipelines to move oil from production areas to refineries, according to a number of studies and publications we reviewed.[26]
- ***Truck:*** According to a 2014 EIA report, U.S. refinery receipts of domestic crude oil by truck increased almost 90 percent from 2008 to 2012, from 69 million barrels to 131 million barrels.[27] In addition, according to a North Dakota Pipeline Authority publication, some natural gas liquids are transported to market by truck.[28]

- ***Barge:*** According to a 2014 EIA report, U.S. refinery receipts of domestic crude oil by barge increased more than 200 percent from 2008 to 2012, from 48 million barrels to 151 million barrels.[29] According to the EIA report, the increase in barge shipments may be partially explained by crude oil being transferred to barges from rail cars for the final leg of some journeys to refineries, particularly on the East Coast and along the Mississippi River.

According to a number of studies and publications that we reviewed, in addition to pipeline capacity limitations, rail, barges, and processing facilities and storage facilities also face limitations. For example, a 2013 industry publication identified a backlog for tank cars, needed to transport oil by rail, in the United States at nearly 60,000—representing over 20 percent of the existing U.S. tank car fleet.[30] In addition, a 2014 Congressional Research Service report states that significant development of loading and unloading facilities could be required if rail is to continue substituting for pipeline capacity.[31] Further, a number of studies and publications identified that oil and gas production in some areas can exceed the capacity to process and store the resources.[32] For example, state officials in North Dakota reported in 2013 that maintaining sufficient natural gas processing capacity is a challenge of increased production.

Transportation Infrastructure Limitations and Related Effects Could Pose Environmental and Safety Risks and Have Economic Implications

A number of studies and publications we reviewed identified environmental and safety risks or economic implications from transportation infrastructure limitations. For example:

Risks to air quality: These risks can be the result of intentional flaring—a process of burning the gas developed along with oil—of associated natural gas that results from limited pipeline infrastructure and of engine exhaust from increased truck and rail traffic.

Oil and natural gas are often found together in the same reservoir. The natural gas produced from oil wells is generally classified as "associated-dissolved," meaning that it is associated with or dissolved in crude oil. In areas where the primary purpose of drilling is to produce oil, operators may flare associated natural gas because no local market exists for the gas and transporting to a market may not be economically feasible. In September 2012, we found that flaring poses a risk to air quality because it emits carbon dioxide—a greenhouse gas linked to climate change—and other air pollutants that can increase ground-level ozone levels and contribute to regional haze.[33] In January 2014, the North Dakota Industrial Commission reported that nearly 30 percent of all natural gas produced in the state is flared. According to a 2013 report from Ceres, flaring in North Dakota in 2012 resulted in greenhouse gas emissions equivalent to adding 1 million cars to the road.[34]

Increased truck and rail traffic associated with the movement of oil from well sites also creates a risk to air quality as engine exhaust, containing air pollutants such as nitrogen oxides and particulate matter that affect public health and the environment is released into the atmosphere.[35] Specifically, the Department of State reported in 2014 that increasing the number of unit trains transporting crude oil could increase greenhouse gases emitted directly from the combustion of diesel fuel in trains[36] and in 2011 we found that trucking freight produces more air pollution than other transportation modes.[37] Air quality may also be degraded as fleets of trucks traveling on newly graded or unpaved roads increase the amount of dust released into the air—which can contribute to the formation of regional haze.[38]

Inherent safety risks: Transporting oil and gas by any means—through pipelines, rail, truck, or barge—poses inherent safety risks. However, in January 2013, we found that pipelines are relatively safe when compared with other modes, such as rail and truck, for transporting hazardous goods because pipelines are mostly underground.[39] For example, we found that large trucks and rail cars transporting hazardous materials, including crude oil and natural gas liquids, resulted in far more fatalities and incidents than pipelines. Specifically, we found that from 2007 to 2011, fatalities averaged about 14 per year for all pipeline incidents reported to PHMSA, including an average of about 2 fatalities per year resulting from incidents on hazardous liquid and natural gas transmission pipelines.[40] In comparison, in 2010, 3,675 fatalities resulted from incidents involving large trucks and 730 additional fatalities resulted from railroad incidents. Therefore, increased transport of oil and gas by rail, truck, or barge could increase safety risks.

According to state officials and several publications we reviewed, increased truck traffic resulting from increased oil and gas production can present hazardous driving conditions—particularly on roads not designed to handle heavy truck traffic.[41] Our analysis of data from PHMSA found that in recent years, the number of reported incidents involving the transport of crude oil by truck in both Texas and North Dakota has increased. Specifically, such incidents increased in Texas from 17 incidents in 2008 to 70 incidents in 2013, and in North Dakota they increased from 1 incident in 2008 to 16 incidents in 2013.

Barge accidents also pose safety risks and can have associated environmental and economic effects. For example, according to the U.S. Coast Guard Polluting Incident Compendium, in 2011, a barge struck a bridge on the Lower Mississippi River causing damage to the barge and a discharge of just over 11,000 gallons of oil.[42] In February 2014, a barge crash resulted in the spilling of about 31,500 gallons of crude oil into the Mississippi River, temporarily shutting down transportation along the river. According to a 2012 Congressional Research Service report, an oil spill from a barge can cause significant harm to marine ecosystems and individual aquatic organisms and negatively affect business activity near the spill, particularly businesses and individuals that count on the resources and reputation of the local environment.[43] For instance, the local fishing and tourist industry may be affected, and in some cases, a well-publicized oil spill can weaken local or regional industries near the spill site, regardless of the actual threat to human health created by the spill.

Economic implications: According to a number of studies and publications we reviewed, infrastructure limitations and related effects could have economic implications, including lost revenue, higher energy prices, and hindered development.

- ***Lost revenue:*** In addition to the risks to air quality from flaring, we found in October 2010 that flaring natural gas has economic implications,[44] and in April 2014 the Environmental Protection Agency reported that flaring results in the destruction of a valuable resource.[45] For example, in 2010 we found that on federal oil and gas leases, natural gas that is flared, instead of captured for sale, represents a loss of about $23 million annually in royalty revenue for the federal government. According to a 2013 report from the North Dakota Pipeline Authority, in August 2013, 2.7 percent of the total economic value and 7.2 percent of the total energy content being produced in North Dakota were lost due to flaring.[46] In another example, a Ceres report found that in May 2013 roughly $3.6 million of revenue was lost per day, at market rates, as a result of flaring in North Dakota.[47]
- ***Higher energy prices:*** Growing shale development and the resulting increased availability and lower prices of natural gas have contributed

to an increasing reliance on natural gas as a source of producing electricity in some parts of the country. However, pipeline infrastructure limitations have at times contributed to price spikes. For example, according to a paper from ICF International, pipeline limitations were a contributing factor to higher natural gas prices in the northeast in January 2014.[48] A cold weather pattern involving record low temperatures led to increased demand for natural gas for space heating and for generating electricity across parts of the country. With the surge in demand, several major pipeline systems became constrained and could not deliver sufficient natural gas to meet demand. According to a 2014 EIA publication, prices at the Algonquin, Massachusetts trading point, which normally are around $3 to $6 per million British thermal units (MMBtu) during unconstrained periods, reached up to $38/MMBtu in early January.[49] These price increases for natural gas led electricity systems to use more oil-fueled generating resources during this period.

- ***Hindered oil and gas production:*** A 2013 study sponsored in part by the Utah Department of Transportation found that oil and gas production from the Uinta Basin is likely to be constrained by limitations in the capacity of transportation infrastructure. Specifically, the study found that existing pipelines in the state are already at or near capacity, and by 2020, demand on the infrastructure network to transport oil and gas will exceed capacity—resulting in a loss of 12 percent of potential production over the next 30 years.[50] Further, according to a 2013 industry report, infrastructure constraints such as pipeline limitations and bottlenecks from the Permian Basin in Texas to a key hub have contributed to discounted prices for some domestic crude oils.[51] For example, we found in March 2014 that West Texas Intermediate crude oil—a domestic crude oil delivered to a key hub that is used as a benchmark for pricing for all crude oil—was priced just under $18 per barrel less in 2012 than Brent, an international benchmark crude oil from the European North Sea that has historically been about the same price as West Texas Intermediate.[52] These discounted prices mean resource developers have received lower prices for their crude oil production. According to a 2013 Energy Policy Research Foundation report, discounted prices may eventually lead to production growth constraints.[53]

DOT HAS NOT FULLY ADDRESSED SAFETY RISKS FROM EXPANSION OF FEDERALLY UNREGULATED GATHERING PIPELINES

Increases in the Number, Size, and Operating Pressure of Gathering Pipelines Pose Additional Risk, Particularly Where They Are Not Federally Regulated

Gathering pipeline construction has increased significantly as a result of increased shale oil and gas development; however, the increase in pipeline mileage is unknown because data on gathering pipelines are not systematically collected by PHMSA nor by every state. The Interstate Natural Gas Association of America (INGAA), a trade organization representing interstate natural gas transmission pipeline companies, estimated in March 2014 that shale oil and gas development will result in approximately 14,000 miles of new gas gathering pipelines and 7,800 miles of new oil gathering pipelines added each year from 2011 through 2035.[54] State officials in Pennsylvania, North Dakota, Texas, and West Virginia said that companies have invested significantly in gathering pipeline infrastructure. For example, according to data published by Texas state officials, 15,684 new miles of federally unregulated gathering pipelines were added in the state between 2010 and 2013.[55] In response to the growth in gathering pipelines, Texas officials told us that their state enacted legislation to increase state regulatory authority over gathering pipelines. Similarly, North Dakota passed rule changes in 2013 to increase state regulatory authority over gathering pipelines. Texas officials told us that they plan to study and determine what parts of their rules should apply to gathering pipelines during 2014 and then issue guidance in 2015. In April 2014, North Dakota implemented regulations requiring companies to report the location and characteristics of gathering pipelines carrying any products including natural gas, crude oil, natural gas liquids, water, and others. The National Association of Pipeline Safety Representatives, an association representing state pipeline safety officials, produced a compendium of state pipeline regulations showing that most states with delegated authority from PHMSA to conduct intrastate inspections do not have expanded regulations that cover increased oversight of gathering pipelines.[56] As a result, companies building gathering pipelines in rural areas are generally not subject to inspection and do not have to report the location and characteristics of much of the gathering pipelines being installed.

Although the majority of the total gathering pipeline network that exists are the traditional small pipelines, state pipeline regulators, PHMSA officials, and pipeline operators we spoke with said that some newly built gathering pipelines have larger diameters and higher operating pressures that more closely resemble transmission pipelines than traditional gathering pipelines. For example, while gathering pipelines have traditionally been 2 to 12 inches in diameter, one company operating in the Texas Eagle Ford shale region showed us plans to build 30- and 36- inch natural gas gathering pipelines, which is near the high end of diameters for regulated transmission pipelines. Historically, federally unregulated gathering pipelines were low pressure, smaller-diameter pipelines and were generally in rural areas where there was less safety risk. Now, according to PHMSA, industry, and state pipeline safety officials we spoke to, gathering pipelines of larger diameter and higher pressure are being constructed, including in areas closer to populations. Such construction could increase safety risk, since an incident occurring on one of these larger, high-pressure unregulated gathering pipelines could affect a greater area and be as serious as an incident involving a regulated transmission pipeline of similar diameter and pressure.

Pipeline operators and industry organizations told us that new gathering pipelines are likely safer because new pipelines are less susceptible to issues like corrosion—a common reason for failure in older pipelines. Pipeline operators also told us that some large-diameter, high-pressure gathering pipelines are built to the same specifications as regulated transmission pipelines and that these pipelines are in very rural areas with little risk to people. They also expressed that safety is very important to the industry and that companies understand not only the potential harm to the network, people, and environment, but also the public perception following a high-profile incident and therefore manage their assets to avoid incidents. Nonetheless, state pipeline regulators, PHMSA officials, and safety organizations expressed concern with the potential safety threat of unregulated gathering pipelines of this size. For example, a citizens' shale development awareness group in Pennsylvania has documented construction of several unregulated gathering pipelines in the state that are 24 inches in diameter. The group argues that while these gathering pipelines are in rural areas, they are being built unnecessarily close to homes. PHMSA officials told us that the large diameter and pressure of the pipelines increase the concern for the safety of the environment and people nearby.

In addition to potential increased safety risk as a result of the changing characteristics of the pipelines, some stakeholders shared concerns about the readiness of emergency responders to address potential incidents that could occur with unregulated gathering pipelines. PHMSA's emergency response planning requirements that apply to other pipelines do not apply to rural unregulated gathering pipelines. Consequently, response planning in rural areas with unregulated gathering lines may be inadequate to address a major incident. Transmission pipeline operators with pipelines similar in size to the new gathering pipelines are required to develop comprehensive emergency response plans and coordinate with local emergency responders. Emergency response officials whom we spoke with stated that lacking information about the location of some gathering pipelines, responders—particularly in rural areas—may not be adequately prepared to respond to an incident. A representative from the National Association of State Fire Marshals told us that training and communication with pipeline companies are key for emergency responders' knowledge and awareness. Additionally, emergency response officials also told us that rural areas in particular lack the level of hazardous-materials response resources found in metropolitan areas where more is known about the extent of local pipeline networks. The National Transportation Safety Board (NTSB) has also stated that emergency response planning is critical for pipeline safety and has recommended that pipeline operators help ensure adequate emergency response by providing local jurisdictions and residents with key information on the pipelines in their areas.

DOT Has Not Proposed Regulatory Changes to Address Risks Posed by Gathering Pipelines

As previously discussed, PHMSA applies a risk-based approach to regulating pipeline safety. A key principle of risk-based management is promoting the use of regulations, policies, and procedures to provide consistency in decision-making. PHMSA has acknowledged the growing potential risk of federally unregulated gathering pipelines as more are constructed and at larger diameters and higher pressures, but DOT has not proposed regulatory changes to address this risk. In August 2011, PHMSA published an Advance Notice of Proposed Rulemaking, stating that the existing regulatory framework for natural gas gathering pipelines may no longer be appropriate due to recent developments in gas production.[57] In the notice, PHMSA asked for comment on whether it should consider establishing new, risk-based safety requirements for large-diameter, high-pressure gas gathering pipelines in rural locations, among other potential changes to gathering pipeline regulations.[58] The proposal also states that enforcement of current requirements has been hampered by the conflicting and ambiguous language of the current regulation that can produce multiple classifications for the same pipeline system, which means that parts of a single pipeline system can be classified as rural gathering pipelines and therefore be unregulated, while other parts of the same pipeline with the same characteristics are regulated. PHMSA officials told us they have drafted proposed regulations for both gas and hazardous liquid gathering pipelines but as of June 2014, the agency had not issued the Notice of Proposed Rulemaking for comment. According to DOT officials, the proposed gas rule is being reviewed internally and the proposed hazardous liquid rule is with the Office of Management and Budget for review.[59] PHMSA officials also told us they have studied existing federal and state gathering pipeline regulation to help identify where gathering pipelines are currently regulated and remaining gaps; however, this study has been in the final stages of review during the course of our work and has not yet been published.

Given the lack of PHMSA regulation of rural gathering pipelines, the extent, location, and construction practices for rural gathering pipelines is largely unknown by federal, state, and local officials, and oversight to verify the construction and monitor operators' safety practices is lacking. In 2012, we concluded that unregulated gathering pipelines also pose risks due to construction quality, maintenance practices, and limited or unknown information on pipeline integrity.[60] We recommended at that time that PHMSA collect data on unregulated gathering pipelines to facilitate quantitatively assessing the safety risks posed by these pipelines, which we said could assist in determining the sufficiency of safety regulations for gathering pipelines. According to DOT officials, as of July 2014, PHMSA has compiled data on existing gathering pipeline requirements, and the resulting report is under internal review. Furthermore, officials said that data collection is part of the proposed rules also under review. In 2010, the National Association of Pipeline Safety Representatives recommended that PHMSA modify federal pipeline regulations to establish requirements for gathering pipelines in rural areas that are presently not regulated. The association stated that with the advent of new production technologies, there has been rapid development of gas production from shale formations such as the Barnett, Marcellus, and Bakken resulting in a significant amount of new gathering pipeline construction. Further, in these newer gas gathering systems, it is not uncommon to find rural gathering pipelines up to 30 inches in diameter and operating at 1480 psi, which is the higher end of traditional transmission operating pressure.[61] Enhanced pipeline safety for all types of pipeline was also on NTSB's "Most Wanted List" in 2013 and 2014. NTSB has prioritized overall pipeline safety because of the increased demand in oil and gas and the aging pipeline infrastructure.

Resources are an important consideration in evaluating how to address the increased risk of gathering pipelines. According to PHMSA officials, inspection resources are limited and would be further stretched if rural gathering pipelines were regulated. If PHMSA were to receive increased staff funding in the near future, there could be a lag in ramping up the inspection workforce because inspectors would have to complete PHMSA's 3-year pipeline-inspection training to become fully certified.[62] However, if PHMSA were to set minimum federal regulations for gathering pipelines, this would enable the agency to include currently federally unregulated rural gathering lines in decisions for prioritizing resources for addressing safety risks. This is in line with the principles of risk-based management, while also enabling data-driven, evidence-based decisions about the risks of rural gathering pipelines, which our previous work has shown is especially important in a time of limited resources.[63] State regulators in all four states we spoke with acknowledged their resources could also be strained; however, the officials supported regulating rural gathering pipelines. State officials said that without rural gathering pipeline regulation, such as provisions for inspections or industry reporting, they have limited knowledge of the construction and maintenance practices of rural gathering pipeline operators, do not always know where new rural gathering pipelines are being constructed, and may not even have communication with the operators.

Transmission Pipeline Infrastructure Has Not Increased Substantially as a Result of Shale Development

Gas flows from shale plays into transmission pipelines have increased, but construction of new transmission pipelines has not increased dramatically as a result of increased shale development. According to PHMSA data, approximately 4,500 miles of new oil and gas transmission pipelines were built between January 2010 and December 2012. This includes about 2,000 miles each of crude oil and natural gas pipelines and the remaining is pipeline for natural gas liquids.

Oil and gas pipeline industry representatives and transmission pipeline companies we spoke with stated that transmission pipeline companies have been able to accommodate increased demand in various ways. Transmission pipeline companies have repurposed existing pipelines, made operational changes—including increased compression and change in directional flow—and added additional capacity to the current network through smaller-scale construction projects. Companies have repurposed pipelines by changing them from one product to another, such as converting a natural gas pipeline to a natural gas liquid pipeline or a crude oil pipeline. Companies have also instituted operational changes such as adding additional compression to their line, which allows them to move more gas through the same lines, and changing directional flow of a pipeline. One pipeline operator we spoke with is reversing flow of its gas transmission pipeline. Prior to shale development, this pipeline moved natural gas from the Gulf Coast to the Northeast. By 2017, the volume of gas that once flowed north will flow south. The pipeline operator stated that changing the direction of the flow, while not as easy as flipping a switch, requires significantly less time and money than building a new pipeline. Smaller-scale construction projects, such as short pipeline extensions, help meet the demands of shale areas like the Marcellus because the natural gas being produced is in close proximity to its destination market. Therefore, it is primarily a matter of connecting new gas production into the existing transmission pipeline network.

However, accommodating increased demand through new construction is a challenging proposition. To build a transmission pipeline requires a long-term commitment—often a contract that spans 30 to 50 years from producers. Major transmission pipeline projects may also face long timeframes. Pipeline companies we spoke with stated that once contracts are in place it usually takes 2 to 4 years to complete a pipeline. This is in part a result of the permitting process, which can include multiple federal and state agencies as well as obtaining rights to build on properties of individual land owners.[64] Timelines can be longer if the pipeline construction project is contentious.[65] The construction timeline is also dependent on terrain and weather. Pipeline industry representatives in West Virginia told us that constructing pipelines in mountainous terrains is much more difficult than in flat land. State officials in North Dakota said that long winters and short construction seasons cause construction projects to last over several construction seasons to complete. Other reasons provided by state officials and industry officials for a slower growth in pipeline infrastructure in North Dakota include the challenge of securing right-of-ways, as well as uncertainty in market demand. It is possible that continued shale exploration and development in other parts of the country could displace demand for Bakken shale products. Developers have proposed some future pipeline projects, and some have been approved in areas like North Dakota, but transmission pipeline mileage has not seen the same kind of rapid increase as gathering pipeline mileage. For example, Enbridge Energy Partners has proposed a transmission pipeline called Sandpiper that spans 612 miles from North Dakota through Minnesota to Wisconsin. The pipeline is proposed to be 24 inches and 30 inches in different places and will carry between 225,000 and 375,000 barrels of oil. WBI Energy announced planning for a 375-mile natural gas transmission line called the Dakota Pipeline in January 2014. The proposed route would start in North Dakota and continue into Minnesota.

In areas of shale development without access to an established pipeline network, such as the Bakken region, lengthy timelines and high costs associated with building transmission pipeline have led producers to seek alternative methods for transporting some of the production—primarily rail.

DOT IS WORKING TO ADDRESS RISKS RELATED TO THE INCREASE IN TRANSPORTATION OF OIL BY RAIL

Crude Oil Is Transported Increasingly by Rail and Travels Long Distances in Large Volumes

The use of rail to transport crude oil from development areas to refineries has increased dramatically. STB data show that rail moved about 236,000 carloads of crude oil in 2012, which is 24 times more than the approximately 9,700 carloads moved in 2008 (see fig. 5).[66] Carloads further increased in 2013, with AAR reporting that Class I freight railroads originated 407,761 carloads of crude oil that year.[67]

According to railroads, the majority of this increased movement of crude oil by rail is done using unit trains, which are trains that carry only one commodity to a single destination. Crude oil unit trains may consist of 80 to 120 tank cars, each carrying about 30,000 gallons of product, for a total of about 2.4 million to 3.6 million gallons of crude oil per train (see figure 6). This has resulted in an increase in demand for tank cars. According to AAR, nearly 50,000 tank cars were used to transport crude oil by rail as of April 2014.[68]

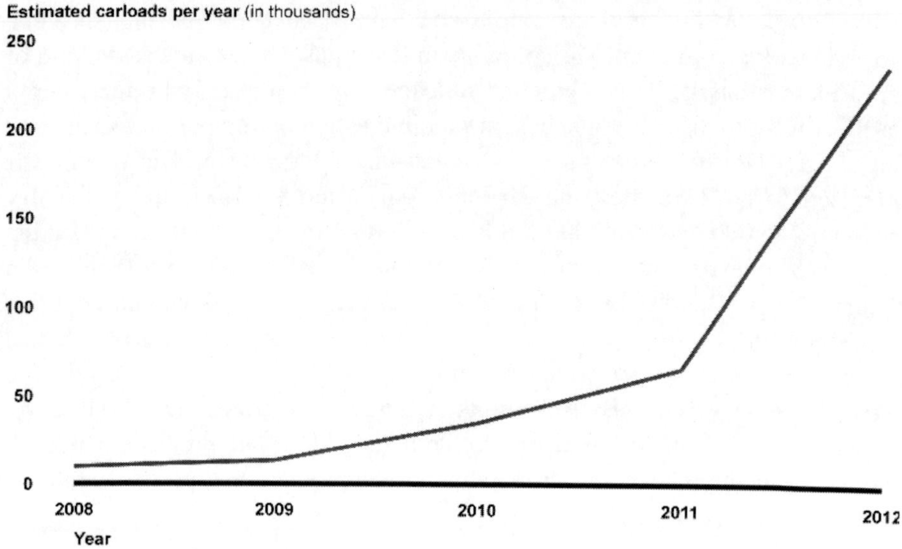

Source: GAO analysis of Surface Transportation Board Data. | GAO-14-667.

Note: The number of crude oil carloads, which was small in 2008-2011, had for those years a declining margin of error that started at 19 percent in 2008 and went down to about 8 percent in 2011. For 2012, the margin of error for crude oil carloads was less than 5 percent.

Figure 5. Estimated Number of Crude Oil Rail Carloads in the United States, 2008 to 2012.

Source: GAO. | GAO-14-667.

Figure 6. Crude Oil Unit Train.

There are different types of crude oil, which affects where crude is refined, as refineries are configured differently to handle the various types. This, in turn, affects where crude oil is transported for refining.[69] Increasingly, crude oil produced in the United States is “light and sweet” (lower in density and sulfur content); in contrast, a portion of new Canadian production has been “heavy and sour” crude oil (higher in density and sulfur content).[70] We have previously reported that not all U.S. refineries can take advantage of domestic crude oils to the same extent because of configuration constraints at some refineries.[71] Therefore, oil may travel long distances to a refinery with the matching refining configuration even if there is refinery capacity nearer to the crude oil source. According to an oil industry association in North Dakota, much of the domestic refining capacity for Bakken crude oil, which is a lighter crude, is located along the Gulf, East, and West Coasts. According to STB data, about 69 percent of crude oil transported by rail in 2012 originated in North Dakota, followed by Texas and all other states (see fig. 7). STB data show that crude oil originating in North Dakota in 2012 traveled to 19 destination American states or Canadian provinces across North America. While most Bakken crude oil was shipped to destinations along the Gulf Coast, there was a large increase in oil shipped to East and West Coast destinations in 2012, signaling a shift in demand from the Gulf region to the other coasts.

While pipelines generally deliver commodities to a fixed customer, rail offers the flexibility to serve different customers, allowing shippers to shift product quickly in response to market needs and price opportunities. Despite the great increase in crude oil transported by rail, the commodity remains a small percentage of railroads’ business, comprising about 1.4 percent of Class I railroads’ freight originations in 2013, according to AAR. As previously discussed, there are thousands of miles of track in the United States, providing various shipping opportunities for crude oil, as well as other commodities.[72] Officials from Class I railroads said they have not extensively added new infrastructure to specifically accommodate the increased shipping of crude oil by rail, although officials from some railroads said they have added track infrastructure in specific areas of increased shale oil development to increase capacity.

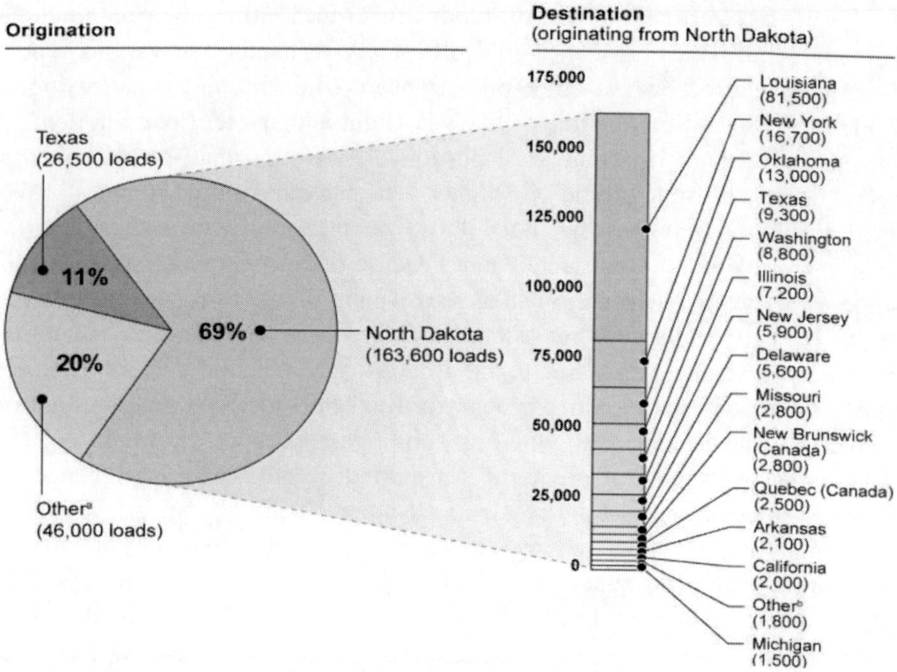

Sources: GAO analysis of Surface Transportation Board data. | GAO-14-667.

[a] "Other" originations include 22 other American states and Canadian provinces, among the largest of which are Saskatchewan, Alberta, Oklahoma, California, and Colorado.

[b] "Other" destinations include 5 other American states and Canadian provinces where the estimated carloads received was fewer than 1,000, including oil that was shipped to final destinations within the state of North Dakota. Multiple destinations may exist in some states.

Notes: Carload numbers are rounded to the nearest hundred. The sum of carloads by destination does not equal the number of carloads originated in North Dakota due to rounding. The margins of error for estimates of carloads received by individual states or provinces are as follows: Missouri (57%), Quebec (55%), New Brunswick (52%), Arkansas (50%), California (40%), Delaware (40%), Michigan (38%), New Jersey (35%), Washington (32%), Texas (27%), Illinois (26%), Oklahoma (25%), New York (22%), and Louisiana (9%). For example, we are 95 percent confident that the value for Missouri is within +/- 57 percent of the estimate itself. Additionally, the relative margins of error for estimates of origination carloads are as follows: Texas (16%), North Dakota (7%), and all other originations (9%).

Figure 7. North American Originations and Destinations of Crude Oil Carloads Hauled by Rail, 2012.

DOT Is Taking Action to Address Safety Risks Resulting from the Transport of Crude Oil by Rail

As the movement of crude oil by rail has increased, incidents, such as spills and fires involving crude oil trains have also increased. PHMSA's hazardous materials incident data show that rail crude oil incidents in the United States increased from 8 incidents in 2008 to 119 incidents in 2013.[73] These data show that the majority of the 2013 incidents were small; however, two incidents in 2013, in Aliceville, Alabama and Casselton, North Dakota resulted in large spills and greater damage. Significant incidents have continued to occur in 2014, including an April derailment and fire in Lynchburg, Virginia. During a presentation at an April 2014 forum on rail safety, NTSB noted that significant accidents involving crude oil have increased in recent years, with one incident occurring between 2008 and 2012 compared to eight incidents since 2012.

DOT, primarily through PHMSA and FRA, sometimes jointly, has taken steps to engage the rail and oil and gas shipping industries and emergency responders to address the safety of transporting crude oil by rail, particularly in response to concerns stemming from the July 2013 Lac-Mégantic, Quebec accident:

- In August 2013, February 2014, and May 2014, DOT issued emergency orders to compel shippers and railroads to address safety risks by taking steps to secure unattended trains, ensure proper testing and packaging of crude oil, and notify emergency responders about crude oil shipments.[74] DOT also issued safety advisories during this period recommending additional actions.[75]
- In August 2013, PHMSA, with FRA assistance, initiated an ongoing special inspection program to examine whether crude oil rail shipments are appropriately tested and packaged. The effort consists of spot inspections, data collection, and testing crude oil samples taken from tank cars. Initial information from the program has identified deficiencies; DOT issued fines against three companies in February 2014 for not following proper crude oil packaging procedures. According to PHMSA officials, this effort inspects about 2 percent of Bakken crude oil trains.
- In September 2013, PHMSA issued an Advance Notice of Proposed Rulemaking seeking comments from industry and other stakeholders on improvements to standards for crude oil rail tank cars.[76] This

action was in response to the railroad industry's 2011 petition for improved standards and recommendations from NTSB.

- In January 2014, PHMSA issued a safety alert notifying the general public, emergency responders, shippers, and carriers that Bakken crude oil may be more flammable than traditional heavy crude oil based on tests associated with PHMSA and FRA's special inspection program. PHMSA said it planned to issue final results of the tests at a later date.
- In February 2014, DOT entered into a voluntary agreement with AAR to improve the safety of moving crude oil by rail including increased track inspections, improved emergency braking capabilities, use of a risk-based routing tool to identifying the safest routes, travel at lower speeds, and emergency response training and planning.
- Also in February 2014, PHMSA officials met with emergency responders and industry groups to discuss training and awareness related to the transport of Bakken crude oil.
- In July 2014, DOT issued an update on PHMSA and FRA's joint special inspection program that includes results to date of their crude oil testing efforts and related discussion of the appropriate packaging of oil tested.

Additionally, in July 2014, PHMSA, in coordination with FRA, proposed new rules that align with key areas of concern cited by stakeholders: crude oil classification, testing, and packaging; crude oil tank car design; and emergency response. Specifically, PHMSA issued a Notice of Proposed Rulemaking seeking comment on proposals for new regulations to lessen the frequency and consequences of train accidents involving large volumes of flammable liquids, such as crude oil. The proposal includes new operational requirements for certain trains transporting a large volume of flammable liquids; revisions to requirements for crude oil classification, testing, and packaging; and improvements in tank car standards.[77] Additionally, PHMSA, also in consultation with FRA, issued an Advance Notice of Proposed Rulemaking seeking comments in response to questions about potential revisions to current regulations for emergency response planning for crude oil transported by rail.[78]

Crude Oil Classification, Testing, and Packaging

PHMSA's current regulations for transporting hazardous materials require that shippers classify and characterize the materials they ship to identify the materials' characteristic properties and select an appropriate shipping package based on those properties.[79] "Classification" refers to identifying a material's hazard class, which could be one or more from the list of nine in PHMSA's hazardous materials regulations, such as a flammable liquid or flammable gas. "Characterization" refers to ascertaining other characteristics of the product to determine its proper packing group, a designation based on risk that identifies acceptable packages.[80] Specifically, PHMSA's regulations classify crude oil as a flammable liquid and offer acceptable tank cars under three packing groups based on characteristics such as the oil's flash point—lowest temperature at which a liquid can vaporize to form an ignitable mixture in air—and boiling point.[81] PHMSA's regulations provide options for the tests shippers may use to determine these characteristics.[82] Crude oil with higher boiling and flash points is considered less risky, since it is less likely to form flammable vapor unless exposed to extreme temperatures, and more approved packaging choices exist for such oil than for oil with lower boiling and flash points that could form ignitable vapor at lower temperatures. Substances that are gases, rather than liquid, at ambient temperatures are even more flammable, and thus more stringent packaging requirements apply to flammable gases than to flammable liquids. In particular, flammable gases must be packaged in pressure tank cars, which provide additional safety in the event of an accident.[83]

According to PHMSA officials, because crude oil is a natural resource, it has greater characteristic variability than a hazardous material manufactured under strict specifications or quality guidelines. Thus, testing may need to be done more frequently to make sure the proper packaging rules are followed, since different rules may apply depending on the characteristics of a particular oil shipment. A review by Canadian transportation safety officials determined that the crude oil involved in the Lac-Mégantic accident was packaged under less stringent packing requirements than those which should have been followed, given the flammability characteristics of the oil involved. Although, as DOT officials pointed out, a different packing group would not have changed the package itself, since the type of tank car involved in the incident, the DOT-111 tank car, is allowed for crude oil transport under all three packing groups.[84]

Stakeholders we spoke to have differing views on the volatility of crude oil from the Bakken region, the area where the most crude oil is being shipped by rail. Some industry stakeholders, including the operators of Bakken crude oil rail terminals, characterized Bakken crude oil as being like any other crude oil produced in the United States, while other stakeholders said it has differences that may make it more volatile. In particular, PHMSA and AAR officials said that Bakken oil has variable composition and may sometimes contain higher than usual levels of dissolved natural gases. According to AAR officials, this can lead to flammable gases building up in a tank car during transport. AAR officials also said that the presence of natural gas makes fires more likely when crude oil tank cars are involved in an accident. Additionally, a May 2014 industry study noted that Bakken crude oil may contain higher amounts of dissolved flammable gases; however, the report states that it is not enough to warrant new regulations for crude oil rail transportation.[85]

According to PHMSA officials, the current regulatory framework for classifying and packaging crude oil for transport by rail may need to be further examined to ascertain if it addresses all of the risks posed by some shale crude oils that have properties unlike other typical crudes. PHMSA's July 2014 Notice of Proposed Rulemaking calls for shippers of mined gases and liquids to be transported by rail, including crude oil, to develop and implement a program for sampling and testing to ensure the shippers' materials are properly classified and characterized. The procedures must outline a frequency of sampling and testing that accounts for the potential variability of the material being tested, sampling at various points to understand the variability of the material during transportation, sampling methods that ensure a representative sample of the entire packaged mixture is collected, and testing methods used, among other requirements. The sampling and testing program must be documented in writing and retained for as long as it remains in effect and be made available to DOT for review upon request.

Representatives from railroads and crude oil terminals we spoke to, as well as from the oil and gas industry, have indicated that clarification about the requirements for testing and packaging crude oil is needed. Specifically, two of the railroads and two crude oil rail terminal operators told us that PHMSA needs to clarify its crude oil testing requirements, including to more clearly state which tests should be done and with what frequency.[86] One of the terminal operators told us that without clearer guidance, they are unsure whether they are performing the right tests and testing with sufficient frequency. They are also concerned they may be incurring unnecessary expense from over-testing. PHMSA's July 2014 Notice of Proposed

Rulemaking does not state which tests should be performed or specifically how often, but does state that testing methods used should enable complete analysis, classification, and characterization of the material as required by PHMSA's regulations and that the frequency of testing should account for the potential variability of the material. The notice also seeks comment on whether more or less specificity of these requirements would aid shippers and whether the proposed guidelines provide sufficient clarity for shippers to understand whether they are in compliance. PHMSA has also drafted more detailed guidance on classification and packaging crude oil, including testing procedures, but had not released it publicly as of June 2014.[87]

Additionally, the American Petroleum Institute, an oil and gas industry association, formed a working group in 2014 to develop industry standards for testing and packaging of crude oil for transportation by rail, which the group hopes to implement by October 2014. PHMSA officials said that PHSMA scientists have been attending the group's meetings and providing input. In its July 2014 Notice of Proposed Rulemaking, PHMSA noted that it is encouraged by the development of an industry standard and that once finalized, PHMSA may consider adoption of such a standard.

Crude Oil Tank Car Design

Under PHMSA's current packaging regulations, a number of types of tank cars are approved for transporting crude oil.[88] However, DOT-111 tank cars are most commonly used, according to industry and railroad representatives, and PHMSA's regulations allow its use for all types of crude oil, regardless of packing group.

NTSB has documented a history of safety concerns with the DOT-111 tank car. Specifically, NTSB has raised concerns regarding the tank car's puncture resistance, heat tolerance, and potential for overpressurization during a fire. In its report of a 2011 investigation of the derailment of a train hauling ethanol tank cars, NTSB noted that its 1991 safety study and four train-derailment investigations from 1992 to 2009 had identified problems with DOT-111 tank cars.[89] The report further concluded that the car's poor performance suggested that DOT-111 tank cars are inadequately designed to prevent punctures and breaches and that the catastrophic release of hazardous materials can be expected when derailments involve DOT-111 cars.[90] In response to that and other rail incidents, in 2012, NTSB recommended that

PHMSA upgrade its DOT-111 tank car standards to improve tank shielding and puncture resistance, a move the industry had already begun to address.

In 2011, the railroad industry petitioned PHMSA to adopt improved standards for DOT-111 tank cars and worked with tank car manufacturers and other stakeholders to develop improved industry standards that were implemented later that year. These standards called for a thicker shell to improve puncture resistance, shielding at both ends of the tank car, and protection for the top fittings of the tank car. More recently in November 2013, following the Lac-Mégantic accident, the railroad industry called for further tank car upgrades, including a thicker shell, protection to prevent overheating, additional shielding, protection for outlet handles on the bottom of a tank, and high-capacity pressure relief valves.[91] Figure 8 shows how these various proposed upgrades may be incorporated into a crude oil tank car.

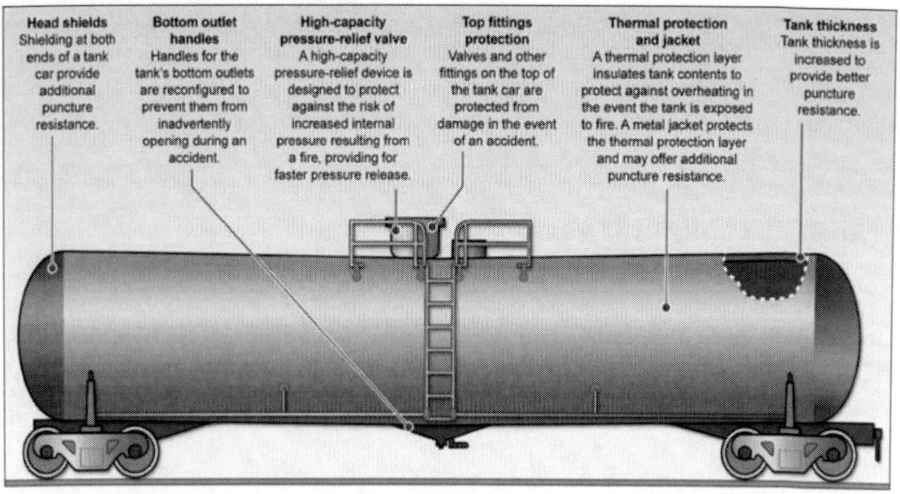

Source: GAO, AAR and railroads. | GAO-14-667.

Figure 8. Proposed Upgrades to Crude Oil Rail Tank Cars.

Although tank cars are generally owned by shippers or third parties, one railroad told us they intend to acquire their own fleet of tank cars built to the railroad industry's 2013 proposed standards. The railroad hopes to incentivize the cars' use by shippers; however, railroads are obligated to move materials as long as they are packaged according to federal standards, and told us they cannot force customers to use upgraded cars and have to accept cargo so long as it is in an allowable package, which includes older model DOT-111 cars.

A wide range of stakeholders we interviewed—including those from PHMSA, NTSB, state transportation agencies, railroad industry, and rail suppliers—told us that crude oil tank car standards need to be improved. Most shippers, railroads, and rail suppliers providing comments in response to PHMSA's September 2013 rail safety Advance Notice of Proposed Rulemaking also stated this opinion. However, there were some differences in their views on how improvements should be implemented. Those who commented in response to PHMSA's notice supported enacting the industry's upgraded tank car standard into regulation and were generally supportive of proposals to strengthen tank cars' puncture resistance through design features such as thicker tank walls, jackets, and shielding. However, stakeholders disagreed on the extent to which existing tank cars should—or even could—be retrofitted to meet higher standards. For example, shippers and rail suppliers stated that existing tank cars built to the current industry standards that already exceed the regulatory standard should be exempt from retrofitting if PHMSA were to adopt an even higher standard. Shippers also expressed concerns that retrofitting would be costly, take tank cars out of service, and put a burden on the already busy shops that also build new tank cars. Local-government and rail-industry commenters supported retrofitting existing cars.

In its July 2014 Notice of Proposed Rulemaking, PHMSA sought comment on requirements for a new DOT-117 tank car standard to replace the current DOT-111 standard for newly manufactured tank cars transporting flammable liquids, which could be one of three options: 1) a design by PHMSA and FRA that would increase puncture resistance, provide thermal protection, protect top fittings and bottom outlets, and improve braking performance; 2) the design in the railroad industry's proposal from November 2013 previously discussed; and 3) the 2011 industry-developed tank car design with some enhancements. PHMSA's Notice of Proposed Rulemaking concluded that cars built to the option 3 standard would likely be built in the absence of a new rule, based on commitments from industry, but that options 1 and 2 would provide additional safety benefits, along with additional cost. Specifically, PHMSA estimated that the improved braking, roll-over protection and increased shell thickness under option 1 would cost $5,000 more per car than option 3. According to PHMSA, option 2 would have most of the same safety features as option 1 except rollover protection and the improved braking system, resulting in a cost of $2,000 more per car than option 3. In addition to the new DOT-117 standard, the proposal would create a performance standard for the design and construction of new tank cars equivalent to the DOT-117, subject to FRA approval. Under the proposal, existing tank cars would have to be retrofitted to comply with the performance standard.[92] The proposal calls for phasing out use of all DOT-111 tank cars for transporting flammable liquids by October 1, 2020, although the cars could still be used to transport other materials.[93]

Although this proposed rule had previously been scheduled for release in November 2014, PHMSA accelerated its efforts to issue a proposal, resulting in the July 2014 release. According to PHMSA officials, the agency did not issue a proposal sooner because the industry, as well as the public, have had different opinions on tank car specifications. Since 2011, PHMSA has received multiple petitions seeking changes to tank car safety standards. In the interim, the railroad industry has moved to adopt higher standards, and in a May 2014 safety advisory, PHMSA and FRA asked companies to refrain from using older tank cars if possible.

Emergency Response

Transporting a large volume of flammable liquid in one train increases the risk of a large fire or explosion in the event of a derailment, such as in the Lac-Mégantic incident. DOT has noted that the transportation of crude oil in unit trains compounds the risk of ignition, and NTSB has reported that crude oil unit trains present the potential for disastrous consequences in the event of an accident. Associations representing emergency responders told us they are particularly concerned about the risk these trains pose to rural areas, which generally have fewer resources to respond to hazardous materials incidents. They also cited concerns with the general lack of awareness about risks and the need for industry to better communicate with local responders about them. Railroad officials told us that risks from unit trains can be managed. Further, railroad officials told us that transporting crude oil in trains that carry a mixture of freight commodities could be higher risk, due to the need to sort crude oil tank cars in rail yards, and that doing so would lead to reduced efficiency by increasing the turn-around times for crude oil trains. In August 2013, AAR revised its guidance on hazardous materials operating practices so that its restrictions would apply to crude oil unit trains; these restrictions include a 50 MPH speed limit, limitations on the use of track siding, and requirements for addressing defective bearings.[94] However, associations representing emergency responders told us that industry should do more to prepare responders for potential incidents, such as by providing information, training, and resources. These organizations also shared concerns about rural responders lacking resources and information to respond as effectively as responders in urban areas. As previously mentioned, stakeholders told us that rural areas may lack sufficient resources to respond to a major event, like an accident involving a crude oil unit train.

As previously discussed, PHMSA has engaged emergency responders on crude oil transportation safety, and the voluntary commitment DOT secured from railroads included emergency response planning and training efforts. However, PHMSA's requirements for comprehensive emergency response planning do not apply to unit trains used to transport crude oil, raising concerns about the abilities of responders and other stakeholders to effectively handle potential incidents. As currently worded, PHMSA's regulations require comprehensive plans for trains that haul any liquid petroleum or non-petroleum oil in a quantity greater than 42,000 gallons per package—greater than the about 30,000 gallons of crude oil typically transported in a tank car—even though a unit train of 100 cars could carry about 3 million gallons of crude oil.[95] Instead, PHMSA requires railroads to have a basic plan that includes information about the maximum potential discharge, response plans, and identification of (but not coordination with) private response personnel and the appropriate people and agencies to contact in the event of an incident.[96] Federal regulations require that comprehensive emergency response plans include a written plan outlining contingency planning, an identified central coordinating official during an incident, private personnel secured by contract or other means to respond to a worst-case incident, training, equipment, and response actions and be subject to review by FRA.[97] Without a comprehensive plan, PHMSA does not have assurance that railroads have taken steps to plan for response needs and identified and coordinated with the appropriate responders. PHMSA's July 2014 Advance Notice of Proposed Rulemaking seeks comment on several possible ways of expanding the comprehensive planning requirement to include crude oil unit trains by changing the threshold under which such plans are required.[98]

Although this review focuses on the packaging and movement of crude oil in tank cars, our prior work has found that while DOT has taken actions in this area, there are other safety issues that are also relevant to the context of the safety of transporting crude oil by rail. Specifically, in a December 2013 report, we found that FRA has developed a risk-based approach to direct its inspection efforts, but the agency had been slow to implement broader risk reduction planning.[99] As required by the Rail Safety Improvement Act of 2008, FRA was tasked with overseeing railroads' development of risk reduction plans.[100] Specifically, FRA was required to issue a final rule by

October 2012 directing railroads to develop these plans, but our report found that FRA had not yet issued the final rule. Our report described safety challenges that railroads face, some of which can contribute to derailments. Other actions have been taken subsequently. FRA issued a final rule in January 2014 revising track inspection requirements to increase the standard for track used to transport hazardous materials.[101] And, as discussed, railroads entered into a voluntary agreement with DOT in February 2014 to improve the safety of crude oil trains.[102] In their comments in response to PHMSA's September 2013 Advance Notice of Proposed Rulemaking on rail safety, several shippers noted that recent incidents have generally been caused by defective track, railroad equipment or operational issues, and supported improvements in these areas. NTSB's accident report for the aforementioned 2011 derailment of ethanol tank cars noted that although problems with tank cars were a contributing factor, the probable cause of the accident was a broken rail. PHMSA has also noted that addressing the causes of derailments, not just upgrading tank cars, is important for improving the safety of transporting crude oil by rail. According to PHMSA officials, the severity of a derailment may present a wide range of forces for any particular tank car to withstand, and therefore, even an enhanced tank car may have variable performance and may not always perform better in a given derailment. PHMSA's July 2014 Notice of Proposed Rulemaking to address safety of transporting crude oil by rail includes a number of other provisions in addition to those already discussed for trains carrying 20 or more tank carloads of flammable liquids, including a routing analysis, enhanced braking, and codifying the May 2014 emergency order requiring notification to emergency responders about crude oil shipments.

CONCLUSION

The advent of new oil and gas production technologies has created a new energy boom for the United States. However, with this increase in production comes the responsibility to move those flammable, hazardous materials safely.

While the Department of Transportation has worked to identify and address risks, its regulation has not kept pace with the changing oil and gas transportation environment. Gathering pipeline construction has increased, but some of these new pipelines in rural areas fall outside the current safety framework, despite operating at the size and pressure (and therefore similar risk) as federally regulated transmission lines. DOT began a rulemaking to address this issue in 2011 but did not issue proposed rules. Subsequently, new gathering pipeline infrastructure has continued to grow, with industry predicting such growth will continue for the foreseeable future, raising concerns where such pipelines are not subject to safety regulations.

The growth in the use of rail to move crude oil has likewise revealed risks not fully addressed by the current safety framework, particularly in ensuring that oil is properly tested and packaged for shipping. Emergency responders also need to be adequately prepared in the event that incidents occur, both for pipeline and for rail. Recent transportation incidents, such as the July 2013 train accident in Lac-Mégantic, Quebec, have highlighted the need for risk-based federal safety oversight. Since the Lac-Mégantic accident, much emphasis has been placed on the need to upgrade standards for tank cars that carry crude oil, but attention to tank cars alone is not sufficient to address safety, a sentiment shared by some railroads and shippers, as well as DOT. Oil and gas shippers, railroads and pipeline operators, emergency responders, and government all have a role to play. Shippers, in particular, play an important role in making sure that hazardous materials like crude oil are properly packaged for safe transport. This underscores the importance of DOT's role in assessing the risk such oil poses and providing clear guidance for handling it safely. Without timely action to address safety risks posed by increased transport of oil and gas by pipeline and rail, additional accidents that could have been prevented or mitigated may endanger the public and call into question the readiness of transportation networks in the new oil and gas environment. DOT's recent proposed rulemakings to address concerns about transporting crude oil by rail signal the department's commitment to addressing these important safety issues. Because of the ongoing rail safety rulemakings, we are not making recommendations related to rail at this time.

RECOMMENDATION

To address the increased risk posed by new gathering pipeline construction in shale development areas, we recommend that the Secretary of Transportation, in conjunction with the Administrator of PHMSA, move forward with a Notice of Proposed Rulemaking to address gathering pipeline safety that addresses the risks of larger-diameter, higher-pressure gathering pipelines, including subjecting such pipelines to emergency response planning requirements that currently do not apply.

Agency Comments and Our Evaluation

We provided a draft of this report to DOT for comment. We received written comments from DOT's Assistant Secretary for Administration. These comments stated that PHMSA generally concurred with our recommendation to move forward with a rulemaking to address risks posed by gathering pipelines. Further, the letter stated that PHMSA is developing a rulemaking to revise its pipeline safety regulations and is examining the need to adopt safety requirements for gas gathering pipelines that are not currently subject to regulations. Additionally, the letter stated that proposed regulations are under development to ensure the safety of natural gas and hazardous liquid gathering pipelines that include collecting new information about gathering pipelines to better understand the risk they pose.

In the version of the draft report we sent to DOT for comment, we had also recommended that PHMSA develop and publish additional guidance on testing, classification and packaging of crude oil for transport by rail and that PHMSA address emergency response planning regulations for transporting oil by rail so that they include shipments of crude oil by unit trains. DOT's written response stated that PHMSA generally concurred with these recommendations and was taking steps to address them. Subsequently, on July 23, 2014, PHMSA, in coordination with FRA, issued rulemaking proposals that, if implemented, would likely address these concerns. Therefore, we are no longer making those recommendations in this report and we have added language to the report describing the objectives of the proposals. We also received technical comments from DOT, which we incorporated as appropriate.

Susan Fleming
Director,
Physical Infrastructure Issues

Frank Rusco
Director,
Natural Resources and Environment

APPENDIX I: OBJECTIVES, SCOPE, AND METHODOLOGY

This report addresses (1) challenges, if any, that increased domestic oil and gas production poses for U.S. transportation infrastructure and examples of associated risks and implications; (2) how pipeline infrastructure has changed as a result of increased oil and gas production, the key related safety risks, and to what extent the U.S. Department of Transportation (DOT) has addressed these risks; and (3) how rail infrastructure has changed as a result of increased oil production, the key related safety risks, and to what extent DOT has addressed these risks.

To identify challenges increased domestic oil and gas production poses for U.S. transportation infrastructure and examples of the associated risks and implications, we reviewed and synthesized information from 36 studies and other publications from federal, state, and tribal government agencies; industry; academics; and other organizations. We identified these studies and publications by conducting a search of web-based databases and resources—including Transport Research International Documentation, ProQuest, and FirstSearch—containing general academic articles, government resources, and "gray literature."[1] Studies and publications were limited to those focused on domestic onshore oil and gas production and published in the years 2008 through 2013. In addition, we reviewed prior work we have conducted. We included examples of known transportation infrastructure limitations and associated effects from these studies and publications in this report. We believe the studies and publications identified through our literature search and included in our review have identified key examples of known transportation infrastructure limitations and associated effects. In addition, we analyzed data from the U.S. Department of Energy's Energy Information Administration (EIA) to identify oil and gas produced from 2007 to 2012. To assess the reliability of these data, we examined EIA's published methodology for collecting this information and found the data sufficiently reliable for the purposes of this report.

To determine how pipeline infrastructure has changed as a result of increased oil and gas production, we analyzed data from DOT's Pipeline and Hazardous Materials Safety Administration (PHMSA) on pipeline construction from January 1, 2010 through December 31, 2012 and interviewed PHMSA officials and representatives of pipeline industry associations and operators. We assessed the reliability of the data on pipeline construction by reviewing documentation about the database, interviewing agency officials about how the data are collected, comparing the data to similar information from EIA on completed pipeline projects, and reviewing the agency's related internal controls. We determined that the data were sufficiently reliable for describing new pipeline construction projects. We identified key pipeline safety risks by reviewing documents provided by and interviewing officials from PHMSA, pipeline industry associations and operators, and safety organizations. To examine trends in pipeline incidents, we analyzed PHMSA's pipeline incident data from January 1, 2008 through December 31, 2013. This analysis only examined transmission pipeline incidents, since many gathering pipelines are not regulated and therefore the data may not include potential gathering pipeline incidents. We assessed the reliability of these data by reviewing documentation on the collection of these data, interviewing agency officials about how the data are collected and whether there are potential limitations for using the data as we intended, and reviewing the agency's related internal controls. We determined that these data were sufficiently reliable for identifying trends in pipeline incidents. We also examined transportation infrastructure changes and safety risks specific to key shale development areas in four states selected because they are located above shale plays in different parts of the country with generally the highest levels of oil and gas production from 2007 through 2011, according to EIA data. The states and corresponding shale plays were Pennsylvania and West Virginia (Marcellus shale play), North Dakota (Bakken shale play), and Texas (Eagle Ford shale play). In these states, we spoke with state oil and gas regulatory and transportation agencies, oil and gas industry associations, oil and gas companies, railroads, and crude oil rail terminal operators, as well as a community advocacy organization in Pennsylvania. We also reviewed documents provided by these organizations.

To determine how rail infrastructure has changed, we analyzed Surface Transportation Board (STB) data on crude oil shipments by rail for calendar years 2008 through 2012 and interviewed DOT officials from PHMSA and the Federal Railroad Administration and industry representatives, including railroads.[2] We assessed the reliability of the STB data by reviewing documentation about the data, interviewing agency officials about how the data were collected and ways they could be analyzed, and reviewing the agency's related internal controls. We determined that the data were sufficiently reliable for describing trends in the movement of crude oil. To identify the key safety risks related to changes in rail infrastructure, we analyzed PHMSA's data on rail hazardous-materials incidents from January 1, 2008 through December 31, 2013, reviewed documents submitted to a DOT rulemaking proceeding on rail safety, and interviewed DOT officials and representatives from safety organizations and industry. We assessed the reliability of PHMSA's incident data by reviewing documentation about the data, interviewing agencies officials about how the data were collected, testing the data for inconsistencies, and reviewing the agency's related internal controls. We concluded that the data were sufficiently reliable for discussing trends in rail hazardous-materials incidents. Additionally, to examine infrastructure impacts and safety issues closely associated with shale areas, we Interviewed officials from state oil and gas regulatory and transportation agencies, industry associations and oil and gas transportation companies in the four states mentioned previously: North Dakota, Pennsylvania, Texas, and West Virginia. To evaluate to what extent DOT has addressed safety risks, we reviewed federal laws and regulations, DOT emergency orders and guidance, and interviewed DOT officials. National and state-level stakeholders we interviewed are listed in tables 2 and 3.[3]

Table 2. National-Level Industry and Safety Stakeholders GAO Interviewed

Stakeholder	Description
Industry stakeholders	
American Petroleum Institute	Oil and gas industry
American Short Line and Regional Railroad Association	Railroad industry
American Trucking Associations	Trucking industry
Association of American Railroads	Railroad industry
Association of Oil Pipe Lines	Pipeline industry
BNSF Railway	Class I railroad
CN	Class I railroad
Canadian Pacific	Class I railroad
CSX Transportation	Class I railroad
Dominion	Gas pipeline operator
Interstate Natural Gas Association of America	Pipeline industry
National Tank Truck Carriers	Trucking industry
Norfolk Southern	Class I railroad
Railway Supply Institute	Tank car manufacturers
Spectra	Gas pipeline operator
Williams/Transco	Gas pipeline operator
Union Pacific Railroad	Class I railroad
Safety stakeholders	

Table 2. (Continued)

Stakeholder	Description
Commercial Vehicle Safety Alliance	Highway safety
International Association of Fire Chiefs	Emergency responders
National Association of Regulatory Utility Commissioners	State regulatory officials
National Association of State Fire Marshals	Emergency responders
National Fire Protection Association	Emergency responders
National Transportation Safety Board	Federal incident investigation
Pipeline Safety Trust	Pipeline safety advocates

Source: GAO. | GAO-14-667.

Table 3. State-Level Stakeholders GAO Interviewed

Stakeholder	Description
North Dakota	
Crestwood Colt Hub	Rail crude oil terminal operator
Hess Corporation	Rail crude oil terminal operator
North Dakota Department of Mineral Resources	State oil and gas regulator
North Dakota Department of Transportation	State highway agency
North Dakota Petroleum Council	Oil and gas industry
North Dakota Pipeline Authority	State pipeline agency
North Dakota Pipeline Association	Pipeline industry
North Dakota Public Service Commission	State oil and gas regulator
Savage Services	Rail crude oil terminal operator
Pennsylvania	
C.O.G.E.N.T.	Community pipeline safety
MarkWest	Gas company
Pennsylvania Department of Transportation	State transportation agency
Pennsylvania Public Utility Commission (Gas Safety Division)	State gas regulator

Stakeholder	Description
Wheeling & Lake Erie Railroad	Regional railroad
Texas	
Gardendale Railroad	Short line railroad
Marathon Oil	Oil and gas pipeline operator
Railroad Commission of Texas	State oil and gas regulator
South Texas Energy Economics Roundtable	Oil and gas industry
Southcross	Gas company
Texas Department of Transportation (Rail Division)	State rail agency
Texas Oil and Gas Association	Oil and gas industry
Texas Pipeline Association	Pipeline industry
West Virginia	
Energy Partners	Gas pipeline operator
West Virginia Department of Transportation (Division of Highways)	State highway agency
West Virginia Public Safety Commission	State oil and gas pipeline regulator

Source: GAO.| GAO-14-667.

We conducted this performance audit from August 2013 to August 2014 in accordance with generally accepted government auditing standards. Those standards require that we plan and perform the audit to obtain sufficient, appropriate evidence to provide a reasonable basis for our findings and conclusions based on our audit objectives. We believe that the evidence obtained provides a reasonable basis for our findings and conclusions based on our audit objectives.

APPENDIX II: IMPACTS OF SHALE OIL AND GAS DEVELOPMENT ON HIGHWAYS IN SELECTED STATES

Pipeline and rail are used to transport shale oil and gas long distances; however, truck transportation via local roads and highways also plays an important role. Trucks transport these goods from production areas to pipeline and rail as well as haul most materials needed to develop oil and gas, such as water, sand, and equipment used during drilling and hydraulic fracturing. State agencies within the four states we examined (Pennsylvania, North Dakota,

Pennsylvania, and Texas) have noted significant local road and highway impacts and safety concerns as a result of shale oil and gas development and have taken steps to address these impacts.

Infrastructure Impacts

States have reported a significant increase in truck traffic as a result of shale oil and gas development, the effects of which are particularly acute in rural areas unused to this level of road congestion. State officials provided estimates of the number of truck-loads required to drill and fracture a shale gas well ranging from about 1,200 to about 3,000. Although the number of trucks is greatest during the initial drilling and fracturing phases, significant truck volume may return if a well is re-fractured and, in the case of oil wells, trucks can be used to remove oil if the wells are not connected by pipeline. State officials told us that roads in many of these areas prior to development were built for light local and farm use and were not built for the additional thousands of heavy truck loads associated with oil and gas development, leading to deterioration. North Dakota officials told us that many of the roads the oil industry is using were built to handle approximately 600 loads a day and now these roads can see thousands of heavy truck loads per day. Officials also told us that state-wide, truck traffic accounts for approximately 18 percent of all traffic, but in development areas truck traffic can account for 35 to 50 percent. As a result, it has been much more difficult to predict where the biggest road deterioration is going to happen because it depends on the location and intensity of shale development. In Pennsylvania, officials told us the increased volume of trucks has shortened the roads' normal life cycle, leading to accelerated deterioration and significant damage. The costs of shoring up and rebuilding roads to address these impacts are significant. The Texas Department of Transportation estimated an annual impact to farm-to-market roads, state highways, and local roads in the Eagle Ford area of about $4 billion.

Safety Concerns

Road deterioration and increased truck volumes have created safety concerns in these states. Reported highway incidents involving crude oil have increased in recent years in North Dakota and Texas, the two states we examined with significant shale oil development.[1] In Texas, for example, the Texas Department of Transportation reported an increase in highway crashes in the Eagle Ford shale and Permian Basin areas, with the Permian area seeing a 13 percent increase in roadway fatalities between 2012 and 2013.

Actions to Address Impacts and Safety

The following actions are examples of ways state officials said they have addressed highway infrastructure and safety concerns:

- *Extraction taxes.* Officials in North Dakota told us the state uses taxes on extracted mineral resources to pay for road improvements. In 2013, the Texas state legislature voted to transfer a portion of the state's oil and gas severance tax to pay for road maintenance, a measure that will go before Texas voters for approval in November 2014.
- *Use agreements.* In Pennsylvania and West Virginia, states have entered into road-use agreements with energy companies making the companies responsible for their damage to the roads and maintaining them. Companies must also pay a bond as part of the agreement. Officials told us these agreements have helped make companies more responsible for their impact. For example, in Pennsylvania, officials told us industry had invested over $750 million in roadway infrastructure improvements.
- *Public awareness.* Texas launched a public education campaign to alert drivers to the need to use caution when driving through energy-related work zones.

Although much of the impact has been on rural, nonfederal roads, the Federal Highway Administration has been involved in helping states to coordinate information sharing. For example, in 2011, the Federal Highway Administration hosted an information-sharing meeting between officials in Pennsylvania and West Virginia, who told us the session was beneficial.

End Notes

[1] We focused our work on pipeline and rail because pipeline is the most used mode for transporting oil and gas products and rail has seen the largest percentage increase in use in recent years.

[2] "Gray literature" publications may include, but are not limited to, the following types of materials: reports (pre-prints, preliminary progress and advanced reports, technical reports, statistical reports, memorandums, state-of-the art reports, market research reports, etc.); theses; conference proceedings; technical specifications and standards; non-commercial translations; bibliographies; technical and commercial documentation; and official documents not published commercially (primarily government reports and documents).

[3] Shale is a sedimentary rock that is predominantly composed of consolidated clay-sized particles.

[4] EIA, "High value of liquids drives U.S. producers to target wet natural gas resources," *Today In Energy* (May 8, 2014).

[5] Petroleum includes crude oil and petroleum products. Petroleum products include gasoline, diesel fuel, heating oil, jet fuel, chemical feedstocks, asphalt, biofuels, and other products.

[6] Natural gas is generally priced and sold in thousand cubic feet (abbreviated Mcf, using the Roman numeral for 1,000). Units of a trillion cubic feet (Tcf) are often used to measure large quantities, as in resources or reserves in the ground, or annual national energy consumption. One Tcf is enough natural gas to heat 15 million homes for 1 year or fuel 12 million natural-gas-fired vehicles for 1 year.

[7] Technically recoverable gas resources are a subset of in-place resources that are producible given available technology. Technically recoverable resources include those that are economically producible and those that are not. Estimates of technically recoverable resources are dynamic, changing to reflect the potential of extraction technology and knowledge about the geology and composition of geologic formations. For additional information on shale oil and gas estimates and methodology to develop the estimates, see GAO, *Oil and Gas: Information on Shale Resources, Development, and Environmental and Public Health Risks*, GAO-12-732 (Washington, D.C.: Sept. 5, 2012).

[8] We use the term transmission pipeline to refer to both hazardous liquids and natural gas pipelines carrying product over long distances to users.

[9] See 49 C.F.R. §179.200. In addition to DOT specification tank cars, there are also tank cars built to specifications set by AAR.

[10] These arrangements, in which states act as "agents" for PHMSA, can cover hazardous liquid pipelines only, gas pipelines only, or both (49 U.S.C. § 60105). States' pipeline safety offices are allowed to issue regulations supplementing or extending federal regulations for intrastate pipelines, but these state regulations must be at least as stringent as the minimum federal regulations (49 U.S.C. § 60104(c)). If a state wants to issue regulations that apply to pipelines that PHMSA does not regulate, such as federally unregulated gathering pipelines, there are no minimum federal standards that need to be adhered to, and the state is free to regulate as it sees fit.

[11] 49 C.F.R. 192, Subpart O.

[12] 49 C.F.R. Part 192.5 and 49 C.F.R. § §195.1(a)((4) and 195.11(a)(2).

[13] For pipelines, there are 48 states, the District of Columbia, and Puerto Rico in PHMSA's natural gas pipeline program and 17 states in its hazardous liquid pipeline program (49 U.S.C. § 60104(c)).

[14] 49 C.F.R. Parts 171-179.

[15] FRA has 30 state partners for rail safety.

[16] 49 C.F.R. § § 192.615 (a)(8) and 195.402 (c)(12). Operators of hazardous liquid pipelines must also establish a response-training program and maintain firefighting equipment for their personnel who will execute the spill response plan, including firefighting equipment, techniques, and use of protective clothing.

[17] 49 C.F.R. § 130.31.

[18] We applied these principles to PHMSA's efforts in our previous report on pipeline safety, see GAO, *Gas Pipeline Safety: Guidance and More Information Needed before Using Risk-Based Reassessment Intervals*, GAO-13-577 (Washington, D.C.: June 27, 2013). A fourth principle, monitoring performance, was also discussed in that report.

[19] IHS Global Inc., *Oil and Natural Gas Transportation & Storage Infrastructure: Status, Trends, & Economic Benefits*, prepared for the American Petroleum Institute (Washington, D.C.: December 2013).

[20] In February 2013, we found that both the interstate and intrastate natural gas pipeline permitting processes are complex and can involve multiple federal, state, and local agencies, as well as public interest groups and citizens, and include multiple steps. According to some industry representatives we spoke with at that time, the interstate permitting process can be time-consuming, depending on the size and complexity of a project. GAO, *Pipeline Permitting: Interstate and Intrastate Natural Gas Permitting Processes Include Multiple Steps, and Time Frames Vary*, GAO-13-221 (Washington, D.C.: Feb. 15, 2013).

[21] According to EIA, a refinery is an installation that manufactures finished petroleum products from crude oil, unfinished oils, natural gas liquids, other hydrocarbons, and oxygenates. Oxygenates are additives, such as ethanol that increase the oxygen content of the fuel.

[22] For example, Fraser Institute, *Intermodal Safety in the Transport of Oil*, Studies in Energy Transportation (October 2013). The Fraser Institute is a public policy research and educational organization. For another example, see John R. Aures and John Mayes, "North American Production Boom Pushes Crude Blending," *Oil and Gas Journal* (May 6, 2013).

[23] GAO, *Petroleum Refining: Industry's Outlook Depends on Market Changes and Key Environmental Regulations*, GAO-14-249 (Washington, D.C.: Mar. 14, 2014).

[24] Kevin Smith, "Risk and Reward from the U.S. Fracking Boom," *International Railway Journal* (Sept. 11, 2013).

[25] EIA, *Annual Refinery Report*, Form EIA-820 (June 25, 2014).

[26] For example, see Fraser Institute, *Intermodal Safety in the Transport of Oil*, Studies in Energy Transportation (October 2013); Paula Dittrick, "US shale production boosts midstream growth," *Oil and Gas Journal* (Nov. 25, 2013); and Statement of Adam Sieminski, Administrator, Energy Information Administrations, U.S. Department of Energy, before the Subcommittee on Energy and Power, Committee on Energy and Commerce, U.S. House of Representatives, 113th Cong., 2nd sess., March 6, 2014.

[27] EIA, *Annual Refinery Report* (2014).

[28] In addition to transporting oil to refineries, in 2012 we found that oil and gas development can require a few hundred to more than a thousand truck loads to transport the water, chemicals, sand, and other equipment needed for drilling and hydraulic fracturing. Further, some of the fracturing fluid that was injected into the well will return to the surface (commonly referred to as "flowback") along with water that occurs naturally in the oil- or gas-bearing formation, collectively referred to as produced water (GAO-12-732). The produced water may also be transported by truck from the well site to an injection well or a wastewater treatment plant.

[29] EIA, *Annual Refinery Report* (2014).

[30] IHS Global Inc., *Oil and Natural Gas Transportation and Storage Infrastructure: Status, Trends, and Economic Benefits* (December 2013).

[31] Congressional Research Service, *U.S. Rail Transportation of Crude Oil: Background and Issues for Congress,* R43390 (Washington, D.C.: Feb. 6, 2014).

[32] For example, Nick Snow, "Massive investment needed for oil, gas facilities, experts say," *Oil and Gas Journal* (May 28, 2012) and Ceres, *Flaring Up: North Dakota Natural Gas Flaring More Than Doubles in Two Years* (July 2013). Ceres is a nonprofit organization.

[33] GAO-12-732.

[34] Ceres, *Flaring Up: North Dakota Natural Gas Flaring More Than Doubles in Two Years* (July 2013).

[35] Nitrogen oxides are regulated pollutants commonly known as NOx that, among other things, contribute to the formation of ozone and have been linked to respiratory illness, decreased lung function, and premature death. Particulate matter is a ubiquitous form of air pollution commonly referred to as soot. GAO, *Diesel Pollution: Fragmented Federal Programs That Reduce Mobile Source Emissions Could Be Improved*, GAO-12-261 (Washington, D.C.: Feb. 7, 2012).

[36] According to the Department of State's Final Environmental Impact Statement for the Keystone XL Pipeline, the use of liquefied natural gas (LNG) as a fuel source for trains is being developed and tested. The use of LNG could reduce greenhouse gas emissions compared to the use of diesel fuel.

[37] GAO, *Surface Freight Transportation: A Comparison of the Costs of Road, Rail, and Waterways Freight Shipments That Are Not Passed on to Consumers*, GAO-11-134 (Washington, D.C.: Jan. 26, 2011).

[38] T. Colborn, C. Kwiatkowski, K. Schultz, and M. Bachran, "Natural Gas Operations From a Public Health Perspective," *International Journal of Human & Ecological Risk Assessment* 17, no. 5 (2011).

[39] GAO, *Pipeline Safety: Better Data and Guidance Needed to Improve Pipeline Operator Incident Response*, GAO-13-168 (Washington, D.C.: Jan. 23, 2013).

[40] More recently, we analyzed PHMSA's pipeline incident data for 2008 through 2013 and found that there was an average of about 3 fatalities per year from incidents involving natural gas pipelines and hazardous liquid pipelines that carry crude oil. Of the 17 reported fatalities during that time, 8 were attributed to a natural gas pipeline incident in 2010.

[41] For example, testimony of Dana "Sam" Buckles Tribal Executive Board Member Assiniboine and Sioux Tribes of the Fort Peck Reservation before the Senate Committee On Indian Affairs Oversight hearing on Tribal Transportation: Pathways to Infrastructure and Economic Development in Indian Country, 113th Cong., 2nd sess., March 13, 2014.

[42] U.S. Coast Guard, *Polluting Incident Compendium Part II* (December 2012).

[43] Congressional Research Service, *Oil Spills in U.S. Coastal Waters: Background and Governance*, RL33705 (Washington, D.C.: Jan. 11, 2012).

[44] GAO, *Federal Oil and Gas Leases: Opportunities Exist to Capture Vented and Flared Natural Gas, Which Would Increase Royalty Payments and Reduce Greenhouse Gases*, GAO-11-34 (Washington, D.C.: Oct. 29, 2010).

[45] Environmental Protection Agency, *Oil and Natural Gas Sector Hydraulically Fractured Oil Well Completions and Associated Gas During Ongoing Production*, (April 2014).

[46] North Dakota Pipeline Authority, "North Dakota Natural Gas: A Detailed Look at Natural Gas Gathering," Oct. 21, 2013.

[47] Ceres, *Flaring Up: North Dakota Natural Gas Flaring More Than Doubles in Two Years* (July 2013).

[48] ICF International is a consulting firm that provides information to public- and private-sector clients.

[49] EIA, *Northeast and Mid-Atlantic power prices react to winter freeze and natural gas constraints* (Jan. 21, 2014).

[50] Duchesne County, Uintah County, Uintah Transportation Special Service District, and the Utah Department of Transportation, *Final Report: Uinta Basin Energy and Transportation Study,* Project No. S-LC47 (14) (Salt Lake City, Utah: April 2013).

[51] John R. Auers and John Mayes, "North American Production Boom Pushes Crude Blending," *Oil and Gas Journal* (May 6, 2013).

[52] GAO, *Petroleum Refining: Industry's Outlook Depends on Market Changes and Key Environmental Regulations*, GAO-14-249 (Washington, D.C.: Mar. 14, 2014).

[53] Energy Policy Research Foundation, *Pipelines, Trains and Trucks: Moving Rising North American Oil Production to Market* (Washington, D.C.: Oct. 21, 2013). The Energy Policy Research Foundation Inc. is a not-for-profit organization that studies energy economics and policy issues.

[54] The INGAA Foundation Inc. *North American Midstream Infrastructure through 2035: Capitalizing on Our Energy Abundance, An INGAA Foundation Report, Prepared by ICF International* (Mar. 18, 2014), available at http://www.ingaa.org/File.aspx?id=21498 (accessed May 16, 2014).

[55] State officials from the other three states could not provide precise numbers on gathering pipeline construction because most of this new construction is in rural locations where pipeline operators are exempt from federal reporting and oversight regulations and these states, unlike Texas, did not have their own reporting requirements. PHMSA regulations for natural gas gathering pipelines do not apply to rural areas designated as Class 1 which is defined as any location with 10 or fewer buildings intended for human occupancy within 220 yards of the centerline of the pipeline for a 1-mile segment of pipeline, see 49 C.F.R. § § 192.5 and 192.8. For hazardous liquid gathering pipelines, PHMSA regulates only certain rural gathering pipelines within one-quarter mile of environmentally sensitive areas, see 49 C.F.R. §195.11. However, as discussed in the background of this report, PHMSA regulates non-rural pipelines.

[56] Based on our analysis, we determined that expanded regulations vary by state, but the compendium shows that at least 6 states have some form of expanded regulation. National Association of Pipeline Safety Representatives, *Compendium of State Pipeline Safety Requirements & Initiatives Providing Increased Public Safety Levels compared to Code of Federal Regulations*, second edition (Sept. 9, 2013).

[57] An Advance Notice of Proposed Rulemaking is an initial step in a rulemaking proceeding seeking comment on issues the agency may address in future proposed regulation.

[58] 76 Fed. Reg. 53086 (Aug. 25, 2011).

[59] 75 Fed. Reg. 63774 (Oct. 18, 2010).

[60] GAO, *Pipeline Safety: Collecting Data and Sharing Information on Federally Unregulated Gathering Pipelines Could Help Enhance Safety*, GAO-12-388 (Washington, D.C.: Mar. 22, 2012). For this report, we surveyed state pipeline agencies, among which there were 39 agencies that reported they had gathering pipelines in their state that PHMSA does not regulate. Eighteen of the state agencies reported that the quality of installation procedures and construction materials is a safety risk for unregulated gathering pipelines because the construction phase of pipeline installation is critical to ensure the long-term integrity of the pipeline. Sixteen state agencies reported that the extent to which pipeline operators maintain their pipelines is a moderate or high safety risk for unregulated gathering pipelines.

According to agency officials, after a pipeline is installed and operational, periodic maintenance—such as inspecting and testing equipment—is important to prevent leaks and ruptures and could extend the operating life of a pipeline.

[61] National Association of Pipeline Safety Representatives, *Urging PHMSA to Establish Regulatory Requirements for Gas Gathering Lines in Class 1 Areas*, Resolution 2010-2 AC-2 (Sept. 30, 2010).

[62] Some state officials also expressed concern about the capacity of PHMSA's inspector training program, which all state inspectors are also required to complete. The 3-year program must be completed mostly in PHMSA's Oklahoma City, Oklahoma, training center, which can create scheduling and budget challenges.

[63] We have applied these criteria to PHMSA before when reviewing its gas integrity management program; see GAO-13-577.

[64] GAO-13-221.

[65] For example, the Keystone pipeline, which would move approximately 100,000 barrels of Bakken crude oil per day, applied for a Presidential Permit in 2008, which has not been approved. For proposed petroleum pipelines that cross international borders of the United States, the President, through Executive Order (EO) 13337, directs the Secretary of State to decide whether a project serves the national interest before granting a Presidential Permit.

[66] Due to a lag in data reporting, 2013 STB data were not available for this report.

[67] We did not independently verify the AAR data.

[68] Specifically, AAR stated that during the period of 2013 through April 2014, there were 104,597 tank cars used to transport flammable liquids by rail, including 49,182 tank cars used for crude oil.

[69] According to DOT officials, commodity price is also a factor in where crude oil is shipped.

[70] Crude oil is generally classified according to two parameters: density and sulfur content. Less dense crude oils are known as "light," while denser crude oils are known as "heavy." Crude oils with relatively low sulfur content are known as "sweet," while crude oils with higher sulfur content are known as "sour." In general, heavier and more sour crude oils require more complex and expensive refineries to process the crude oil into usable products.

[71] GAO-14-249.

[72] We have ongoing work examining freight congestion in railroad corridors and the impact on local communities.

[73] PHMSA's incident database can be searched at https://hazmatonline.phmsa.dot.gov/IncidentReportsSearch/ (accessed June 12, 2014).

[74] FRA's August 2013 emergency order required railroads to take additional steps to secure unattended trains to prevent unintended movement; DOT's February 2014 emergency order, amended in March 2014, specified requirements for testing and classifying crude oil for transportation by rail and required shippers to package crude oil under the more stringent standards as allowed by regulation; DOT's May 2014 emergency order required railroads shipping 1,000,000 or more gallons of crude oil in a single train (about 35 car loads) to notify states' emergency response commissions about the movement of such trains through their states and which specific counties trains would be traveling through.

[75] DOT's August 2013 safety advisory recommended that railroads review the circumstances of the Lac-Mégantic accident and consider changes to their operating practices related to crew staffing, human factors, and train securement. DOT's November 2013 safety advisory reinforced the importance of proper testing and packaging for flammable liquids transported by train. DOT's May 2014 safety advisory asked companies to use tank cars with the

highest integrity available for transporting Bakken crude oil and to avoid using older DOT-111 tank cars.

[76] 78 Fed. Reg. 54849 (Sept. 6, 2013).

[77] 79 Fed. Reg. 45016 (Aug. 1, 2014).

[78] 79 Fed. Reg. 45079 (Aug. 1, 2014).

[79] 49 C.F.R. Parts 172 and 173.

[80] 49 C.F.R. § 172.101. Hazardous liquids, including crude oil, can be packaged under three sets of increasingly stringent requirements known as "packing groups," which range from the least restrictive group III to the most restrictive group I. In addition to specifying what types of tank cars may be used under each packing group, packing groups I and II require the shipper to develop a security plan that includes assessment of transportation security risks and measures to address those risks, see 49 C.F.R. § 172.800 (a)(6). According to PHMSA officials, the security plan would include planning for leaving a train unattended, which could have been a factor in the Lac-Mégantic accident, since that train was unattended at the time of the accident. The crude oil in the Lac-Mégantic accident had been assigned to packing group III, which does not require this plan. Subsequently, investigators have determined that it should have been assigned to packing group II, which would have required a plan. Details of the Lac-Mégantic accident are incomplete because the Transportation Safety Board of Canada's investigation of the accident is still ongoing. As of January 2014, the agency was completing its examination and analysis phase and planned to soon move onto its reporting phase.

[81] 49 C.F.R. § § 172.101(g), 173.120, 173.121, 173.242, 173.243, and 179.200.

[82] 49 C.F.R. § § 173.120(c) and 173.121 (a)(1).

[83] 49 C.F.R § § 172.101, 173.115(a), 173.314, and 179.100.

[84] Transportation Safety Board of Canada, *Rail Safety Advisory Letter—12/13* (Gatineau, Quebec, Canada: Sept. 11, 2013).

[85] American Fuel & Petrochemical Manufacturers and Dangerous Goods Transport Consulting Inc., *A Survey of Bakken Crude Oil Characteristics Assembled For the U.S. Department of Transportation* (May 14, 2014).

[86] The other railroads did not mention the issue, and the third terminal operator said the current guidance is sufficient.

[87] PHMSA provided us a draft of this guidance in March 2014; according to PHMSA officials, the document is still under development.

[88] 49 C.F.R. § § 173.242(a) and 173.243(a),

[89] NTSB, *Railroad Accident Brief, NTSB/RAB-13/02* (Washington, D.C.: Aug.14, 2013).

[90] According to FRA officials, tank car specifications and design requirements were based on forces and stresses incident to normal transportation, not crash survivability, since derailments are not normal transportation events. Officials said that recently crash survivability has been a metric for tank car specification and design.

[91] Additionally, an AAR official speaking at an NTSB forum and officials from one railroad that we spoke to suggested that the type of tank cars currently used may not be appropriate for transporting crude oil of higher flammability and that pressure tank cars may be preferable. PHMSA's hazardous materials regulations require flammable gases to be transported in pressure tank cars. According to FRA officials, if the vapor pressure of crude oil is high enough, regulations would require putting it in a pressure car. According to PHMSA's and FRA's July 2014 update on its special inspection effort, the current classification of Bakken crude oil as a flammable liquid is accurate under the current classification system; however, the update also notes that Bakken crude oil tested has a higher gas content, higher vapor

pressure, lower flash point, and lower boiling point than most other crude oil in the United States. Consequently, the update concluded that Bakken crude oil has a higher degree of volatility than other oil, which correlates to increased ignitability and flammability.

[92] PHMSA's Notice of Proposed Rulemaking notes that retrofitting could be costly, potentially exceeding $30,000 per car, and would result in out-of-service time of at least 1 month. Due to cost concerns, PHMSA's proposal for retrofitting existing tank cars does not include requirements for top-fittings protections.

[93] Specifically, PHMSA's Notice of Proposed Rulemaking states that existing DOT-111 tank cars may be retrofitted to DOT-117 standards, retired, repurposed, or operated under speed restrictions.

[94] AAR's hazardous materials operating practices guidance had previously applied to trains carrying poison inhalation hazards, toxic inhalation hazards, spent nuclear fuel, and high-level radioactive waste, but was expanded to also include trains carrying 20 car loads or intermodal portable tank loads of any combination of hazardous material, which would include unit trains carrying 20 or more carloads of crude oil.

[95] The House Appropriations Committee called on PHMSA to address this problem in a report that accompanied the committee's Transportation-HUD appropriations bill H.R. 4745, for fiscal year 2015. H. Rpt. 113-464, p.62.

[96] 49 C.F.R. § 130.31(a).

[97] 49 C.F.R. § 130.31(b).

[98] Specifically, the notice asks whether the threshold should be set at 1) 1,000,000 gallons or more of crude oil per train, 2) 20 or more carloads of crude oil per train, 3) 42,000 gallons of crude oil per train, or 4) another threshold. The notice states that comments will inform a potential future Notice of Proposed Rulemaking on the issue.

[99] GAO, *Rail Safety: Improved Human Capital Planning Could Address Emerging Safety Oversight Challenges*, GAO-14-85 (Washington, D.C.: Dec. 9, 2013).

[100] Pub. L. No. 110-432, § 103, 122 Stat. 4848, 4853.

[101] 79 Fed. Reg. 4234, 4245 (Jan. 24, 2014).

[102] Additionally, railroads we spoke to said they have invested heavily in upgrading their track in recent years.

End Notes for Appendix I

[1] "Gray literature" publications may include, but are not limited to, the following types of materials: reports (pre-prints, preliminary progress and advanced reports, technical reports, statistical reports, memorandums, state-of-the art reports, market research reports, etc.), theses, conference proceedings, technical specifications and standards, non-commercial translations, bibliographies, technical and commercial documentation, and official documents not published commercially (primarily government reports and documents).

[2] In addition to DOT officials from PHMSA and FRA, we also interviewed officials from the Federal Highway Administration, Federal Motor Carriers Safety Administration, and Maritime Administration. Since we focused our evaluation of DOT's efforts on pipeline and rail, we did not evaluate the efforts of these other DOT administrations.

[3] Additionally, Kinder-Morgan, an oil pipeline operator, declined to participate, and Enbridge, an oil pipeline company, was not able to schedule an interview with us within our time frame.

We also contacted five other crude oil rail terminals in North Dakota that either did not respond to our requests or declined to be interviewed.

End Note for Appendix II

[1] The other two states, Pennsylvania and West Virginia, have shale formations where natural gas is primarily developed, which is not transported by truck.

In: Oil and Gas Transportation
Editor: Elton Simmons

ISBN: 978-1-63463-841-8

Chapter 2

U.S. RAIL TRANSPORTATION OF CRUDE OIL: BACKGROUND AND ISSUES FOR CONGRESS*

John Frittelli, Paul W. Parfomak, Jonathan L. Ramseur, Anthony Andrews, Robert Pirog and Michael Ratner

SUMMARY

North America is experiencing a boom in crude oil supply, primarily due to growing production in the Canadian oil sands and the recent expansion of shale oil production from the Bakken fields in North Dakota and Montana as well as the Eagle Ford and Permian Basins in Texas. Taken together, these new supplies are fundamentally changing the U.S. oil supply-demand balance. The United States now meets 66% of its crude oil demand from production in North America, displacing imports from overseas and positioning the United States to have excess oil and refined products supplies in some regions.

The rapid expansion of North American oil production has led to significant challenges in transporting crudes efficiently and safely to domestic markets—principally refineries—using the nation's legacy pipeline infrastructure. In the face of continued uncertainty about the prospects for additional pipeline capacity, and as a quicker, more flexible alternative to new pipeline projects, North American crude oil producers

* This is an edited, reformatted and augmented version of a Congressional Research Service publication, No. R43390, dated May 5, 2014.

are increasingly turning to rail as a means of transporting crude supplies to U.S. markets. According to rail industry officials, U.S. freight railroads are estimated to have carried 434,000 carloads of crude oil in 2013 (roughly equivalent to 300 million barrels), compared to 9,500 carloads in 2008. In 2014, 650,000 carloads of crude oil are expected to be carried. Crude imports by rail from Canada have increased more than 20-fold since 2011. The amount of oil transported by rail may also be influenced by a tight market for U.S.-built tankers.

While oil by rail has demonstrated benefits with respect to the efficient movement of oil from producing regions to market hubs, it has also raised significant concerns about transportation safety and potential impacts to the environment. The most recent data available indicate that railroads consistently spill less crude oil per ton-mile transported than other modes of land transportation. Nonetheless, safety and environmental concerns have been underscored by a series of major accidents across North America involving crude oil transportation by rail—including a catastrophic fire that caused numerous fatalities and destroyed much of Lac Mégantic, Quebec, in 2013. Following that event, the U.S. Department of Transportation issued a safety alert warning that the type of crude oil being transported from the Bakken region may be more flammable than traditional heavy crude oil.

Legislation introduced in Congress following the Lac Mégantic disaster would require railroads to have at least two crew members aboard all trains. In addition, policy makers are discussing regulatory changes involving tank car design, prevention of derailments, and selection of preferred routes for transporting oil by rail. Congress may evaluate these changes in the reauthorization of the Rail Safety Improvement Act of 2008 (P.L. 110-432).

INTRODUCTION

North America is experiencing a boom in crude oil supply, primarily due to the growth of heavy crude production in the Canadian oil sands[1] and the recent expansion of shale oil production in North Dakota, Montana, and Texas. North American production now supplies 66% of U.S. crude oil demand, displacing crude from Latin America, Africa, and the Middle East.

This shift has led to significant challenges in transportation, as refineries that once received crude oil principally from oceangoing tankers are now seeing increasing deliveries by domestic transport. Existing pipeline capacity is, in some cases, insufficient to carry growing crude oil from some production areas, or does not link to the refineries needing the oil. The domestic barge network does not serve some key production regions located far from navigable waterways. As a quicker, more flexible alternative to new pipeline projects, North American crude oil producers are increasingly turning to rail as a means of transporting crude supplies to U.S. markets. Increased exports of refined products—and, if Congress changes the law, of crude oil—could lead to even larger volumes of oil being transported by rail. According to rail industry officials, U.S. freight railroads are estimated to have carried 434,000 carloads of crude oil in 2013, or roughly 300 million barrels, compared to 9,500 carloads in 2008.[2] In 2014, 650,000 carloads of crude oil are expected.[3] Crude imports by rail from Canada have increased more than 20-fold since 2011.

The rapid increase in crude oil shipments by rail will likely increase the number of oil spills from rail transportation. However, the most recent data available indicate that railroads consistently spill less crude oil per ton-mile transported than other modes of land transportation. The amount of crude spilled per ton-mile of rail transport declined significantly between the early 1990s and the 2002-2007 period, the most recent years for which data are available.[4]

Nonetheless, the increase in rail shipments of crude has raised safety and environmental concerns. These concerns have been underscored by a series of major incidents involving crude oil transportation by rail, including a catastrophic fire and explosion in Lac Mégantic, Quebec, in July 2013 and a derailment in Casselton, ND, in December 2013 that led to a mass evacuation. Consequently, government agencies in the United States and Canada have issued new regulations and are considering others related to oil transport by rail, and some Members of Congress have called for tighter rules governing crude oil railcars as well as a broader reconsideration of the role of rail in the nation's oil transportation infrastructure.[5]

WHY IS OIL MOVING BY RAIL?

In 2013, the United States produced 2.72 billion barrels of crude oil and imported another 2.82 billion barrels.[6] Canada has become the United States' leading foreign supplier, thanks to its increasing production from oil sands.[7] However, U.S. oil output has been increasing rapidly. In October 2013, U.S crude oil production exceeded imports for the first time since February 1995.[8]

The location of U.S. crude oil production has been changing rapidly. In particular, production in Alaska and from offshore sites has been declining, while production in Texas and North Dakota has been rising. The U.S. Geological Survey recently estimated that 2.7 billion barrels of light sweet crude oil remain in overlooked producing formations,[9] including the Eagle Ford shale, a prolific source of very light sweet crude oil in Texas, and the Bakken formation in North Dakota, a source of light sweet crude oil that rivals West Texas crude in quality.[10]

Almost all oil produced domestically, as well as some Canadian production, flows to one of the 115 U.S. refineries (**Figure 1**).[11] Nearly 45% of the country's refining capacity is located in the Gulf Coast, where 43 refineries process more than 9 million barrels of oil per day (bpd). However, the Midwest and the West Coast also have significant refining capacity.

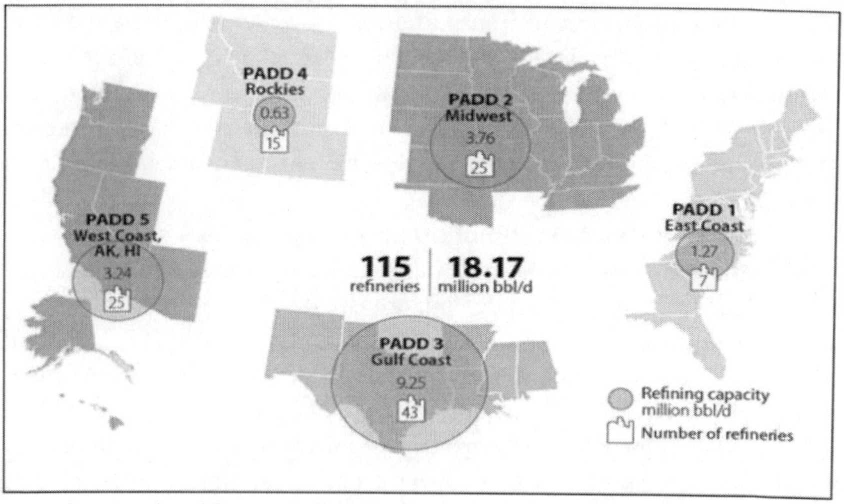

Sources: Congressional Research Service; Energy Information Administration.
Note: PADD = Petroleum Administration for Defense Districts, five districts established by executive order during World War II for gasoline rationing.

Figure 1. U.S. Refinery Capacity by PADD in 2012.

The last entirely new petroleum refinery in the United States opened in 1976. The number of refineries in operation has steadily declined since then as refining capacity has become concentrated in ever larger refineries. A quarter of U.S. capacity is concentrated in 11 refineries with capacities exceeding 300,000 bpd. The largest, Shell/Motiva's Baytown, TX, refinery, was recently expanded to 600,000 bpd. Operable U.S. refining capacity has actually increased from 16.5 million to nearly 18 million bpd over the last decade. Refineries representing approximately 75% of domestic capacity (13.3 million bpd) have the ability to process heavy crude oils, but many smaller refineries can process only light to intermediate crude oil.

Each refinery depends upon a certain grade or blend of crude oils to operate efficiently, depending upon its custom-designed processing equipment. A refinery designed to run light crude oil could not switch to heavy crude oil without adding a coking unit, for example. However some refineries that process heavy sour crude could switch to lighter sweet crude by bypassing their coking units, if the economics of doing so are favorable. Until quite recently, the supply of light sweet crude oil was diminishing, but newly available light sweet crudes from North Dakota's Bakken formation are changing refining dynamics in some regions of the United States, especially as refineries seek supplies that cannot be delivered economically by tanker ships or pipelines.

Traditionally, pipelines and oceangoing tankers have delivered the vast majority of crude to U.S. refineries, accounting for approximately 93% of total receipts (in barrels) in 2012. Although other modes of transportation—rail, barge, and truck—have accounted for a relatively minor portion of crude oil shipments, volumes have been rising very rapidly. The volume of crude oil carried by rail increased 423% between 2011 and 2012, and the volume moving by barge, on inland waterways as well as along intracoastal routes, increased by 53%. The volume of crude oil shipped by truck rose 38% between 2011 and 2012. **Figure 2** shows the change in transportation by mode between 2008 and 2012.

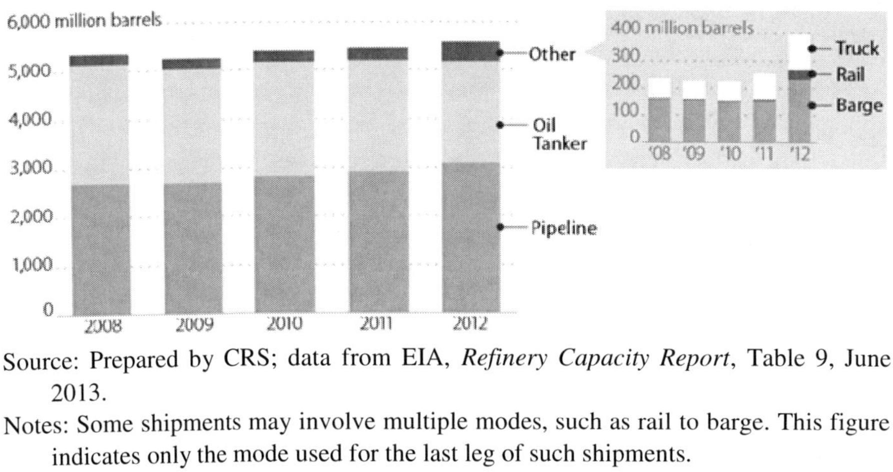

Source: Prepared by CRS; data from EIA, *Refinery Capacity Report*, Table 9, June 2013.

Notes: Some shipments may involve multiple modes, such as rail to barge. This figure indicates only the mode used for the last leg of such shipments.

Figure 2. U.S. Refinery Receipts of Crude Oil by Mode of Transportation.

Rail is a relatively high-cost method of transporting oil. Although crude oil transportation costs are typically not a major driver of refiner profitability, refiners are typically wary of incurring any costs that are higher than those faced by their competitors, as all refined petroleum products sold in a region tend to command the same price independent of the refinery that produced them.

The Economics of Oil by Rail

In the short run, rapid expansion of oil production in the Bakken—production volumes increased nearly ten-fold between 2005 and 2013[12]—strained the capacity of existing pipelines and of refiners able to process the oil. Finding ready buyers was difficult, resulting in discounted prices compared to other crude oil traded in the U.S. market. With Bakken crude selling for approximately $4 to $28 per barrel less than West Texas Intermediate (WTI) crude, the U.S. reference price for crude grade, refiners found it profitable to utilize the North Dakota oil delivered by rail even though the rail transportation cost is perhaps $5 to $10 per barrel higher than pipeline costs.

Rail has also been critical to development of Canadian oil sands. Although the vast majority of crude oil imports from Canada are delivered via existing pipeline, imports by rail are estimated to have increased from 1.6 million barrels in 2011 to 40 million barrels in 2013. Construction of the proposed Keystone XL pipeline could move a significant proportion of these shipments off the rails, as pipeline transportation is likely to cost less per barrel.[13]

For certain refiners, the economics of using rail to transport Bakken oil supplies are even more attractive. In 2012, several refineries in the Philadelphia area were scheduled for closure. The refineries were using imported crudes, largely sourced from West Africa, which sold at a premium to WTI,[14] making their refined products, notably gasoline, uncompetitive against similar products produced by Gulf Coast refineries that used cheaper heavy crudes. By using supplies from the Bakken, these refineries have lowered their costs and have become more competitive. New owners are now investing in the refineries, including installation of high-speed rail unloaders that would allow them to use 230,000 barrels per day of Bakken crude oil by early 2014.[15] These innovations would also reduce the cost of rail transportation per barrel.

The attractiveness of rail transportation of oil may be temporary. Transporting Bakken crude by rail became cost-effective because of the price discounts created by pipeline bottlenecks. If additional oil pipeline capacity were constructed, say from North Dakota to the East Coast market, refiners would likely prefer lower-cost pipeline transportation. And if the refineries could obtain Bakken crude by pipeline, demand would increase, likely reducing or eliminating the current price discount. Without the price discount, Bakken oil would not be competitive in refining when transported by rail. On the other hand, a rising Bakken crude oil price would likely lead to greater drilling activity in the Bakken fields. Given the uncertainty about the future value of the oil and the longevity of the deposits, it is not certain that investors will undertake construction of pipelines from the Bakken fields to the East Coast. In that case, large volumes of crude could be transported by rail well into the future.

Railroads are a viable alternative to pipeline transportation largely because they offer greater flexibility. The nation's railroad network is more geographically extensive than the oil pipeline network, and better able to ship crude oil from new areas of production to North American refineries. While there are about 57,000 miles of crude oil pipeline in the United States, there are nearly 140,000 miles of railroad.[16]

The U.S. Railroad Industry in Brief

The U.S. rail network comprises seven large (Class I) railroads, which focus on moving products between North American regions. These railroads generally market to large volume, long-distance shippers. There are also roughly 500 "shortline" (Class II or III) railroads that sometimes serve as the first or final leg of a Class I rail shipment. Shortlines were often spun-off from Class I railroads because of insufficient business over the line. Class I railroads account for about 70% of system mileage, 90% of railroad employees, and about 95% of freight railroad revenue. Since crude oil movements involve non-traditional rail origins (drilling sites) and destinations (refineries), shortlines are often involved in these movements.

Railroad track is categorized into classes that determine the allowable speeds over the track.[17] Most track with the lowest speed limits is the property of shortlines. If track needs maintenance work, a railroad will issue a "slow order" on that section of track, reducing train speeds.

Class I railroads have transitioned to using bigger and heavier cars, raising the maximum weight on many track sections from 263,000 lbs. to 286,000 lbs. Shortline railroads that interchange traffic with Class I railroads have had to improve their roadbeds to accommodate the heavier cars.

The railroad industry, since 1980, is mostly economically deregulated. The Surface Transportation Board can review the reasonableness of railroad rates and service in situations where the railroad is determined to have "market dominance," generally where a shipper is served by only one railroad and cannot ship economically by other means. As "common carriers," railroads are required to provide rail service upon reasonable request. Railroads do not require a special federal permit to transport crude oil. Federal railroad law preempts state and local authority, which is generally restricted to a state or local government's "police powers."

The geographic flexibility of the railroad network compared to the oil pipeline network can be especially beneficial for a domestic market in flux. Railroads can increase capacity relatively cheaply and quickly by upgrading tracks and roadbeds to accommodate higher train speeds, building passing sidings or parallel tracks, increasing the frequency of switchovers from one track to the other, and upgrading signal systems to reduce the headway needed between trains. Although railroads need approval from the federal Surface Transportation Board (STB) to build new lines, they do not require STB approval to make improvements to existing lines. And even without capacity improvements, railroads can offer routings not served by pipelines.

A significant fall-off in railroad coal movements has increased railroads' capacity to transport oil over some routes. In 2013, railroads carried about 395,000 more tank cars of crude than in 2005, but about 1.3 million fewer cars of coal. To put the increase in crude traffic in perspective, crude oil represented less than 1% of total rail carloads in 2012. In the first three quarters of 2013, crude carloads increased to 1.4% of total rail car loadings. While, on a national scale, increased rail car loadings of crude oil represent a relatively small percentage of total traffic, significant increases in traffic in a specific area can cause bottlenecks that can reverberate across the entire rail network. The STB held a hearing in April 2014 to hear complaints from non-oil shippers concerning poor rail service in the upper Midwest due to oil traffic and the severe winter weather.[18] The STB ordered BNSF and CP railroads to report how they intended to ensure delivery of fertilizer to farmers in spring

2014. At the hearing, BNSF (the railroad most directly serving the Bakken region) noted that its car loadings in North Dakota had more than doubled from 2009 to 2013, and that in October 2013, crude oil and agricultural car loadings surged by more than it could manage. Past experience has shown that railroad bottlenecks are not quickly resolved.

This experience illustrates that pipelines can generally provide more reliable service than railroads. Among other differences, rail shipments are more affected by weather. In addition, railroads generally experience peak demand during the fall due to the grain harvest and retailers' holiday shipments. This may cause locomotives and track capacity to be in shorter supply at certain times of the year.

Railroad transport reportedly costs in the neighborhood of $10 to $15 per barrel compared with $5 per barrel for pipeline. In return, railroads offer oil producers certain advantages. Heated railroad tank cars improve the viscosity of oil sands crude so that less diluent needs to be added than if the product were being moved by pipeline. Generally, railroads are more willing to enter into shorter-term contracts with shippers than pipelines (1 to 2 years versus 10 to 15 years), offering more flexibility in a rapidly changing oil market. Moving oil by train from North Dakota to the Gulf Coast or Atlantic Coast requires about 5 to 7 days' transit, versus about 40 days for oil moving by pipeline, reducing producers' need for working capital to cover the cost of oil in transit.[19]

Crude oil often moves by unit train, a train that carries just one type of cargo in a single type of car and serving a single destination. Unit trains do not need to be switched or shunted in rail yards, saving time and reducing costs, and return to their point of origin as soon as they have been unloaded. A train consisting of 70 to 120 tank cars can carry in the neighborhood of 50,000 to 90,000 barrels of oil, depending on the type of crude.

One hindrance to the expansion of crude-by-rail has been the lack of tank cars and loading and unloading infrastructure. Much of this investment is being made by the oil industry or by rail equipment leasing companies, not railroads. As of April 2014, manufacturers had 50,000 crude oil tank cars on order, on top of an existing fleet of 43,000. (This is in addition to 30,000 tank cars that carry ethanol and 27,000 that carry other flammable liquids.) In 2013, over 28,000 tank cars of all types were built, up from over 17,000 in 2012.[20] Facilities for building tank cars are unique because the process involves baking the entire car in an oven. One manufacturer believes the tank car builders are capable of increasing production each year by 7,000 to 10,000 cars.

Rail terminal capacity is expected to increase fourfold from 2012 to 2015.[21] Matching the daily throughput volume of a pipeline requires several trains per day, with each train taking 13 to 24 hours to unload; oil rail terminals therefore require large areas for parallel loop tracks where multiple trains can await unloading.

The Role of Barges and Ships in Domestic Crude Transportation

Many refineries traditionally have received crude from overseas and thus are located near the coastline with access to dock facilities. Some are not equipped to receive crude by rail. Hence, some railroads are transferring oil to barges for the last leg of the trip to refineries. Locations where railroads transfer crude oil to barges include St. Louis and Hayti, MO; Osceola, AR; Hennepin, IL; Albany, NY; Yorktown, VA; and Anacortes and Vancouver, WA. In addition, crude produced at Eagle Ford, TX, which is located near ports, is being moved along the coast by either barge or ship.

One river barge can hold 10,000 to 30,000 barrels of oil. Two to three river barges are typically tied together in a single tow that carries 20,000 to 90,000 barrels, about the same load as a unit train. Coastal tank barges designed for open seas, known as articulated tug-barges, or ATBs,[22] can hold 50,000 to 185,000 barrels, although newer ATBs can carry as much as 340,000 barrels, comparable to the capacity of coastal tankers. ATBs are slower, less fuel-efficient, and more restricted by sea conditions, but nevertheless may have an economic advantage over tankers because Coast Guard crewing regulations allow them to sail with one-third to half the crew required on a tanker. Crude oil tankers used to move Alaska oil to West Coast refineries have capacities of 800,000 to over 1 million barrels.

An advantage of tankers over railroads is the greater amount of oil they can carry in a single voyage, which better matches the daily consumption rate of refineries. With the median capacity for U.S. refineries at about 160,000 barrels per day, a coastal tanker can carry a two-day supply of oil. In addition, while railroads must build and maintain tracks and pay property taxes on their rights-of-way, the ocean is free, and harbor channels are largely provided by the federal government. For these reasons, tankers can be much cheaper than railroads in moving oil, even though the railroad route may be much more direct. For instance, the distance between the Bakken region in North Dakota and refineries in the Northeast is approximately 1,800 miles, and the cost of railroad transport is $14 per barrel.[23] The distance from Texas ports near the

Eagle Ford region to the same refineries is about 2,100 miles, and tanker rates are $5 to $6 per barrel.[24] Similarly, the overland distance from the Eagle Ford region to Los Angeles-area refineries is about 1,400 miles, and the estimated cost of railroad transport is $15 per barrel, while the water route through the Panama Canal is 5,200 miles and is estimated to cost $10 per barrel.[25]

Although seemingly a circuitous route compared to rail, it is not inconceivable that tankers could play a role in moving Bakken oil to East or West Coast refineries. Significant amounts of Bakken oil are moved to Gulf Coast terminals by pipeline, railroad, barge, or combinations of these modes for refining within that region. From a Gulf Coast port, tankers could transport the oil to either East or West Coast refineries. Via existing rail and pipeline connections to Great Lakes ports, tankers could also move Bakken oil from there to Northeast refineries. However, the economic viability of these routes, in particular, and routes involving domestic coastal transport in general, is heavily influenced by the Jones Act.[26]

The Jones Act

The Jones Act may have a profound impact on where crude oil is sourced and how it is transported. The Jones Act requires that vessels transporting cargo between two U.S. points be built in the United States, as well as crewed and at least 75% owned by U.S. citizens.[27] The domestic build requirement for tanker ships, in particular, has been identified as contributing to higher costs in moving domestic crude oil along the coasts.[28] Domestically built tankers are about four times the price of foreign-built tankers,[29] and there is limited capacity in U.S. shipyards to build them. Much of the existing crude oil tanker fleet was built since 2000 to meet Oil Pollution Act of 1990 (P.L. 101-380) requirements that tankers calling at U.S. ports have double hulls. Two crude carriers are expected to be delivered in 2014 to replace two vessels in Alaska trade.

As of June 2013, the Jones Act-eligible fleet of crude oil tankers consisted of 10 ships, all employed in moving Alaska crude oil to the U.S. West Coast or to a refinery in Alaska.[30] Since annual Alaska oil production has fallen by about 46% over the last decade, the Jones Act crude oil fleet has been in decline. About 30 Jones Act-eligible tankers carry chemicals or refined petroleum products, such as gasoline or jet fuel, but these ships do not readily alternate between carrying dirty oil (crude oil, residual fuel oil, asphalt) and refined (clean) petroleum products because the tanks would have to be extensively washed after carrying dirty product, a time-consuming and costly process. Some product vessels have fundamentally different designs from crude carriers and would require a layup in a shipyard to be converted to move crude oil.

Phillips 66 has chartered two Jones Act tankers to move crude oil from Eagle Ford, TX, to a refinery in Linden, NJ (in proximity to New York Harbor).[31] Phillips 66 has stated that if more Jones Act-eligible tankers were available, it would like to receive 100,000 barrels a day of Eagle Ford oil at this refinery (it would need several tankers to accomplish this, the exact number depending on the size of the tankers).[32] EIA data (which specify oil movement only between regions, not to individual refineries) indicate that over 13 months from January 2013 to the end of January 2014, an average of 22,000 barrels a day of Texas oil (8.5 million barrels total) were shipped by vessel to all U.S. Northeast refineries.[33] Meanwhile, over the same 13 months, twice as much Texas oil (17.1 million barrels) was shipped to refineries in eastern Canada, in foreign-flag tankers. The oil shipments from Texas to Canada cost approximately $2 per barrel, compared with $5 to $6 per barrel for shipments from Texas to U.S. Northeast refineries in Jones Act-qualified tankers. The cost difference for a 300,000-barrel tanker amounts to around $1 million, meaning that a Texas oil producer receives $1 million less for its oil when shipping to U.S. Northeast refineries than when shipping to Canadian refineries.[34] The Bayway refinery also receives 50,000 barrels of Bakken oil per day by rail, and is finishing construction of a rail unloading terminal with capacity of 75,000 barrels per day. The refinery has a capacity of 238,000 barrels per day. Its remaining sources of oil may be offshore oil from eastern Canada (with shipping rates fluctuating around $1.20 per barrel) and Nigeria (with shipping rates around $1.60 per barrel), as EIA data indicate these were the top two sources of foreign oil for Northeast refineries in 2013.[35]

The Role of Tank Trucks

Tank trucks operating on U.S. roadways have been an important link in moving crude oil from domestic drilling sites to pipelines and rail terminals. A typical tank truck can hold 200 to 250 barrels of crude oil. Trucks readily serve the need for gathering product, as the hydraulic fracturing method of drilling employed in tight oil production involves multiple drilling sites in an area and the location of active wells is constantly in flux. A large volume of crude oil is being transported by truck between production areas and refineries in Texas because of the close proximity of the two.

OIL SPILL CONCERNS

Each mode of oil transportation—pipelines, vessels, rail, and tanker trucks—involves some risk of oil spills. Over the period 1996-2007, railroads consistently spilled less crude oil per ton-mile than trucks or pipelines. Barges and domestic tanker ships have much lower spillage rates than trains (**Figure 3**). However, the data in **Figure 3** precede the recent dramatic increase in oil transportation by rail.

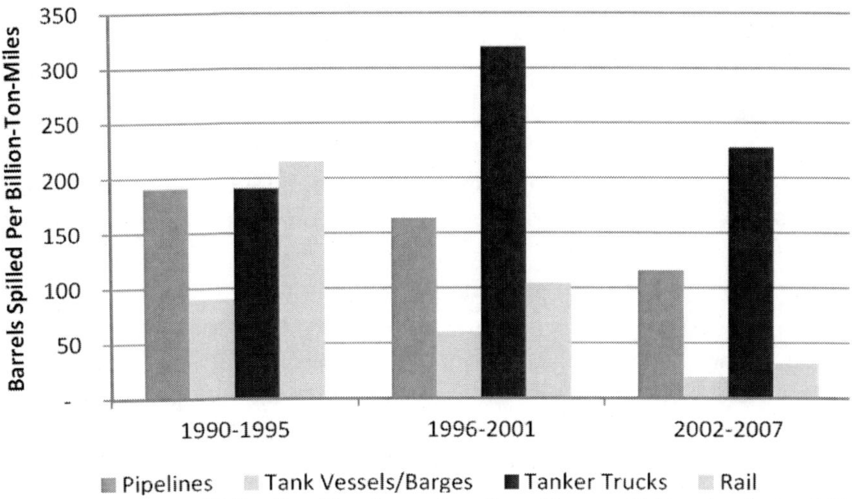

Sources: Prepared by CRS; oil spill volume data from Dagmar Etkin, *Analysis of U.S. Oil Spillage*, API Publication 356, August 2009; ton-mile data from Association of Oil Pipelines, *Report on Shifts in Petroleum Transportation: 1990-2009*, February 2012.

Notes: Pipelines include onshore and offshore pipelines. The time periods were chosen based on the available annual data for both spill volume and ton-miles. The values for each time period are averages of annual data for each six-year period.

Figure 3. Oil Spill Volume per Billion-Ton-Miles; Crude Oil and Petroleum Products during Domestic Transportation.

Given the comparatively small capacity of a rail tank car, around 700 barrels, the total amount spilled from even a major derailment is likely to be small compared to the 260,000 barrels discharged in the 1989 grounding of the *Exxon Valdez* in Prince William Sound, AK, or the approximately 40,000 barrels discharged in the largest U.S. pipeline oil spill CRS can document, which occurred in 1991 near Grand Rapids, MN.[36] Nonetheless, spill volume is arguably a relatively unimportant factor in terms of impacts and cleanup costs. Location matters more: a major spill away from shore will likely cost considerably less to abate than a minor spill in a populated location or sensitive ecosystem. Depending on timing and location, even a small spill can cause significant harm to individual organisms and entire populations.[37]

Although spillage per ton-mile of oil transported by rail declined over time, a recent series of major accidents (see the text box) has heightened concern about the risks involved in shipping crude by rail.

Oil by Rail Derailments in 2013 and 2014

Lac Mégantic, Quebec—On July 5, 2013, a train with 72 loaded tank cars of crude oil from North Dakota moving from Montreal, Quebec, to St. John, New Brunswick, stopped at Nantes, Quebec, at 11:00 pm. The operator and sole railroad employee aboard the train secured it and departed, leaving the train on shortline track with a descending grade of about 1.2%. At about 1:00 AM, it appears the train began rolling down the descending grade toward the town of Lac-Mégantic, about 30 miles from the U.S. border. Near the center of town, 63 tank cars derailed, resulting in multiple explosions and subsequent fires. There were 47 fatalities and extensive damage to the town. 2,000 people were evacuated. The initial determination was that the braking force applied to the train was insufficient to hold it on the 1.2% grade and that the crude oil released was more volatile than expected.

Gainford, Alberta—On October 19, 2013, nine tank cars of propane and four tank cars of crude oil from Canada derailed as a Canadian National train was entering a siding at 22 miles per hour. About 100 residents were evacuated. Three of the propane cars burned, but the tank cars carrying oil were pushed away and did not burn. No one was injured or killed. The cause of the derailment is under investigation.

Aliceville, Alabama—On November 8, 2013, a train hauling 90 cars of crude oil from North Dakota to a refinery near Mobile, AL, derailed on a section of track through a wetland near Aliceville, AL. Thirty tank cars derailed and some dozen of these burned. No one was injured or killed. The derailment occurred on a shortline railroad's track that had been inspected a few days earlier. The train was travelling under the speed limit for this track. The cause of the derailment is under investigation.

Casselton, North Dakota—On December 30, 2013, an eastbound BNSF Railway train hauling 106 tank cars of crude oil struck a westbound train carrying grain that shortly before had derailed onto the eastbound track. Some 34 cars from both trains derailed, including 20 cars carrying crude, which exploded and burned for over 24 hours. About 1,400 residents of Casselton were evacuated but no injuries were reported. The cause of the derailments and subsequent fire is under investigation.

Plaster Rock, New Brunswick—On January 7, 2014, 17 cars of a mixed train hauling crude oil, propane, and other goods derailed likely due to a sudden wheel or axle failure. Five tank cars carrying crude oil caught fire and exploded. The train reportedly was delivering crude from Manitoba and Alberta to the Irving Oil refinery in Saint John, New Brunswick. About 45 homes were evacuated but no injuries were reported.

Philadelphia, Pennsylvania—On January 20, 2014, 7 cars of a 101-car CSX train, including 6 carrying crude oil, derailed on a bridge over the Schuylkill River. No injuries and no leakage were reported, but press photographs showed two cars, one a tanker, leaning over the river.

Vandergrift, Pennsylvania—On February 13, 2014, 21 tank cars of a 120-car train derailed outside Pittsburgh. Nineteen of the derailed cars were carrying crude oil from western Canada, and four of them released product. There was no fire or injuries.

Lynchburg, Virginia—On April 30, 2014, 15 cars in a crude oil train derailed in the downtown area of this city. Three cars caught fire, and some cars derailed into a river along the tracks. The immediate area surrounding the derailment was evacuated. No injuries were reported.

In March and April 2013, there were two derailments of Canadian Pacific trains, one in western Minnesota and the other in Ontario, Canada; less than a tank car of oil leaked in each derailment and neither incident caused a fire.

The increasing deployment of unit trains changes the risks involved in shipping oil by rail in two ways. Unit trains of crude oil concentrate a large amount of potentially environmentally harmful and flammable material, increasing the probability that, should an accident occur, large fires and explosions could result. This risk is similar to that of unit trains carrying ethanol, and maybe greater than that of mixed freight trains in which various hazardous materials, such as explosives and toxic-by-inhalation materials, are sequenced among other cars according to federal regulations.[38] On the other hand, while unit trains concentrate a voluminous quantity of potentially dangerous material, they may offer safety benefits from avoiding the decoupling and re-coupling of cars in rail yards, which involve high-impact forces and introduce opportunity for human error.

Special Concerns About Canadian Dilbit

Oil companies generate substantial quantities of crude oil and related substances from the natural bitumen in oil sands, particularly deposits in Alberta, Canada. In 2012, the United States imported 438 million barrels of oil sands-derived crude oils, 125% more than in 2005.[39] Because bitumen is highly viscous, it is transported mostly in the form of diluted bitumen, or dilbit, containing naptha or other materials that make it flow more easily.

Some commenters have argued that due to its physical characteristics, dilbit presents greater risks of oil spills than conventional crude, with potentially greater impacts to the environment.[40] Other stakeholders and organizations have questioned these conclusions.[41] A study released by the National Research Council in 2013, conducted at the direction of Congress,[42] found that the characteristics of dilbit do not increase the likelihood of spills.[43] The extent to which these findings are applicable to rail transport of crude is open to debate, as rail tanker cars may have different operating parameters (e.g., temperature) and physical standards (e.g., wall thickness), or may transport different forms of oil sands-derived crude oil, decreasing the relevance of the NRC findings.

However, observations in the aftermath of a 2010 pipeline spill are consistent with the assertion that dilbit may pose different hazards, and possibly different risks, than other forms of crude oil. On July 26, 2010, a pipeline owned by Enbridge Inc. released approximately 850,000 gallons of dilbit into Talmadge Creek, a waterway that flows into the Kalamazoo River in Michigan.[44] Three years after the spill, response activities continued,[45] because, according to EPA, the oil sands crude "will not appreciably biodegrade."[46] The dilbit sank to the river bottom, where it mixed with sediment, and EPA has ordered Enbridge to dredge the river to remove the oiled sediment.[47] As a result of this order, Enbridge estimated in September 2013 its response costs would be approximately $1.035 billion,[48] which is substantially higher than the average cost of cleaning up a similar amount of conventional oil.[49]

Special Concerns About Bakken Crude

The properties of Bakken shale oil are highly variable, even within the same oil field. In general, however, Bakken crude oil is much more volatile than other types of crude.[50] Its higher volatility may have important safety implications.

In January 2014, the Pipeline and Hazardous Materials Safety Administration (PHMSA) within the U.S. Department of Transportation (DOT) issued a safety alert warning that recent derailments and resulting fires indicate that crude oil being transported from the Bakken region may be more flammable than traditional heavy crude oil.[51] PHMSA, whose rules are enforced by the Federal Railroad Administration with respect to railroads, reinforced the requirement to properly test, characterize, classify, and where appropriate sufficiently degasify hazardous materials prior to and during transportation. Under its initiative "Operation Classification," PHMSA is to continue to collect samples and measure the characteristics of Bakken crude as well as oil from other locations.

Federal Oversight of Oil Transport by Rail

The Federal Railroad Administration (FRA) has jurisdiction over railroad safety. It has about 400 federal inspectors throughout the country and also utilizes state railroad safety inspectors. State inspectors predominantly enforce federal requirements because federal rail safety law preempts state law, and federal law is pervasive. The FRA uses past incident data to determine where its inspection activity should be targeted, although the FRA Administrator stated that in light of the growth of crude-by-rail transportation, the agency also must look for "pockets of risk."[52] FRA regulations cover the safety of track, grade crossings, rail equipment, operating practices, and movement of hazardous materials (hazmat). The Pipeline and Hazardous Materials Safety Administration within DOT (PHMSA) issues requirements for the safe transport of hazmat by all modes of transportation, which the FRA enforces with respect to railroads.[53]

Rail incidents are investigated by the National Transportation Safety Board (NTSB), an independent federal agency. The NTSB makes recommendations toward preventing future incidents based on its findings.[54] Unlike the FRA, the NTSB is not required to weigh the costs against the benefits when considering additional safety measures and it has no regulatory authority. Many of the NTSB's recommendations concerning oil transport by rail are identical to those it previously issued for transporting ethanol by rail. While the FRA has largely agreed with NTSB's recommendations, its rulemaking process involves consultation with industry advisory committees, and it must determine which of the many rail safety measures under evaluation deserve priority. Implementing a change in FRA regulations can take years.

U.S. safety requirements apply to any train operating in the United States, regardless of its origin or destination. Canadian safety regulations are very similar but do not exactly mirror U.S. requirements. Cross-border shipments must meet the requirements of both countries. Safety standards established by the rail industry, which often exceed government requirements, apply to both U.S. and Canadian railroads.

When a rail incident results in the release of oil, state, territorial, or local officials are typically the first government representatives to arrive at the scene and initiate immediate safety measures to protect the public. The National Oil and Hazardous Substances Pollution Contingency Plan, often referred to as the National Contingency Plan (NCP), indicates that state, territorial, or local officials may be responsible for conducting evacuations of affected populations. These first responders also may notify the National Response Center to elevate an incident for federal involvement, at which point the coordinating framework of the NCP would be applied.

Unlike most federal emergency response plans, which are administrative mechanisms, the NCP is codified in federal regulation and is binding and enforceable.[55] The NCP regulations apply to applicable spills from vessels, pipelines, onshore facilities, and offshore facilities. The definition of "onshore facility" includes, but is not limited to "motor vehicles and rolling stock."[56]

If an oil discharge affects navigable waterways, shorelines, or "natural resources belonging to, appertaining to, or under the exclusive management authority of the United States,"[57] Section 311 of the Clean Water Act, as amended by the Oil Pollution Act of 1990, Section 311(c), provides explicit federal authority to respond.[58] The term "discharge" is defined broadly and is not linked to specific sources of oil. The President has the authority to perform cleanup immediately using federal resources, monitor the response efforts of the spiller, or direct the spiller's cleanup activities.[59] Several executive orders have delegated the President's response authority to the Environmental Protection Agency (EPA) within the "inland zone" and to the U.S. Coast Guard within the coastal zone, unless the two agencies agree otherwise.[60] The lead federal agency serves as the On-Scene Coordinator to direct the federal resources used in a federal response.

Regulations require that railroads have either a so-called “basic” response plan or a more “comprehensive” response plan, depending on the volume capacity of the rail car transporting the oil.[61] Comprehensive plans are subject to FRA approval, and must ensure by contract or other means that personnel and equipment are able to handle a worst-case discharge.[62] However, the regulatory threshold for the comprehensive response plan is a tank car holding more than 1,000 barrels, so does not apply to the DOT-111 tank cars used today, which hold around 700 barrels of oil apiece. For these smaller tank cars, railroads must prepare only “basic” response plans, which are not subject to FRA approval.

This threshold was established in 1996,[63] before the advent of oil unit trains, each of which may transport, in aggregate, approximately 70,000 barrels (almost 3 million gallons) of oil. The NTSB recently recommended that the threshold for comprehensive plans be lowered to take into account the use of unit trains.[64]

Issues for Congress

While oil by rail has demonstrated benefits with respect to the efficient movement of oil from producing regions to market hubs, the dramatic increase in oil by rail shipments has generated interest in several related issues. These include railroad safety,[65] environmental concerns, and trade-offs over rail versus pipeline development.

Railroad Safety and Incident Response

Prior to the Lac Mégantic derailment, the FRA had increased its inspection activity with regard to trains carrying crude oil. After the derailment, the FRA and PHMSA (along with Transport Canada) initiated a comprehensive review of safety requirements.[66] Three areas of active discussion involve tank car design, prevention of derailments, and railroad operations. Railroads are an integrated system of fixed infrastructure, rolling equipment, and workers. Railroad safety experts note that improving safety performance requires recognition of the parameters and interactive effects among these components, and thus approaching railroad safety as an optimization problem is appropriate.[67] For instance, each additional safety feature on tank cars may increase their weight. This reduces the amount of product carried in each car due to track and bridge weight limits, potentially requiring additional tank cars to carry the same amount of product, and thus increasing the expected number of accidents.

In February 2014, the U.S. DOT reached an agreement with railroads under which they would voluntarily take measures to improve the safety of oil trains.[68] These measures include adding braking power; reducing train speeds to 40 mph through urban areas starting July 1 for trains with at least 20 tank cars of crude oil and at least one tank car of the older DOT-111 standard; installing additional wayside wheel bearing detectors by July 1; and other actions. Some of the measures the railroads agreed to are similar to those that the industry already takes for trains carrying “toxic-by-inhalation” hazardous materials. Reducing train speed can reduce the number of cars that derail, as well as the likelihood that product will be released from those tank cars.[69]

Tank Car Safety Design

The U.S. DOT establishes construction standards for tank cars.[70] A tank car used for oil transport is roughly 60 feet long, about 11 feet wide, and 16 feet high (see **Figure 4**). It weighs 80,000 pounds empty and 286,000 pounds when full. It can hold about 30,000 gallons or 715 barrels of oil, depending on the oil’s density. The tank is made of steel plate, 7/16 of an inch thick (see 49 C.F.R. §179.201).[71] An oil tank car is typically loaded from the top valve and unloaded from the bottom valve. Loading or unloading each car may take several hours, but multiple cars in a train can be loaded or unloaded simultaneously.

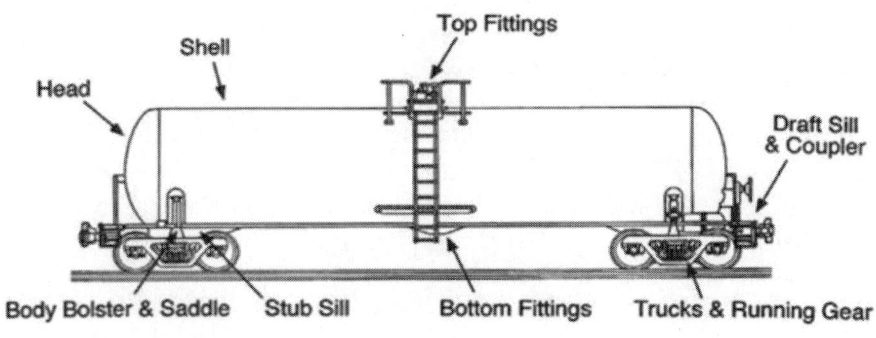

Figure 4. Non-jacketed, Non-pressure Tank Car.

In some incidents, oil has been released from tank cars because the coupler from a neighboring car punctured the tank during derailment. Valves at the top and bottom of the cars have also been sheared off or otherwise opened during derailment. Efforts to improve crashworthiness have focused on reinforcing the shells of tank cars at both ends and/or along the sides with a "jacket" of steel, adding protective head shields at both ends, modifying couplers to prevent decoupling, adding skid protection or diversion shields to protruding valves and modifying pressure relief valves, eliminating or modifying bottom valves so that handles break off rather than opening the valve during derailment, and increasing insulation for fire protection.[72]

The FRA and PHMSA have questioned whether Bakken crude oil, given its characteristics, would more properly be carried in tank cars that have additional safety features, such as those found on pressurized tank cars used for hauling explosive liquids.[73] The railroad industry established additional standards in October 2011 for newly built cars that address some but not all of the safety features that the FRA and PHMSA are considering.

Rail cars have an economic life of 30 to 40 years, so conversion of the fleet to a new car standard could take some time. DOT has asked for further information on the costs and benefits of retrofitting the existing fleet.[74] In November 2013, the Association of American Railroads stated it supports either retrofitting or phasing out oil tank cars built before October 2011 (a fleet of about 78,000 cars) and modifying those built after October 2011 (about 14,000 cars).[75] Some Members of Congress have urged DOT to expedite the rulemaking process concerning new tank car safety requirements.[76]

In February 2014, BNSF announced that it would purchase 5,000 tank cars with safety features that surpass those specified in the October 2011 industry standard (shell thickness would be 9/16 of an inch thick, and they would be jacketed). In April 2014, Transport Canada announced that the oldest DOT-111 tank cars (about 5,000 that lack bottom reinforcement) would no longer be allowed for use in transporting dangerous goods, and the remaining fleet would either be phased out or retrofitted within three years.[77] Transport Canada expects to finalize regulations by the summer of 2014. Reportedly, the U.S. DOT submitted its proposed regulations on tank car safety design to the Office of Management and Budget for review at the end of April 2014.[78] Final U.S. regulations are expected in the summer of 2014, and expected to be in harmony with Canadian regulations. Railroads, shippers, and tank car builders could issue their own standard, but they have not reached agreement on all safety features to be required.

Preventing Derailments

An analysis of freight train derailments from 2001 to 2010 on Class I railroads' mainline track found that broken rails or track welds were the leading cause of derailments, by far.[79] These problems caused 670 derailments over the period, while the next leading problem (track geometry defects) caused just over 300 derailments. Broken rails or welds also resulted in more severe incidents, derailing an average of 13 railroad cars instead of 8.6 cars for all other causes. Broken rails or welds accounted for 23% of total cars derailed. A separate study covering the same time period found that track problems were the most important causes of derailments, followed by problems with train equipment.[80]

In the Rail Safety Improvement Act of 2008 (P.L. 110-432, Section 403(a)), Congress requested that the FRA study and consider revising the frequency and methods of track inspection. The FRA conducted the study and on January 24, 2014 issued a final rule on improving rail integrity.[81] The new rule requires railroads to achieve a specified track failure rate rather than scheduling inspections based on the calendar or traffic volume. It also allows railroads to maximize use of rail inspection vehicle time by prioritizing remedial action when track defects are detected. While the rule requires railroads to achieve a lower track failure rate for track that has higher speed limits, or carries passengers or hazardous material, it does not require lower failure rates for track travelling through populated areas or otherwise consider the affected area around the track in the event of a derailment.[82]

The final rule states that it "codifies standard industry good practices," and notes that railroads "were already initiating and implementing the development of a performance-based risk management concept for determination of rail inspection frequency," meaning that railroads generally have been testing more frequently than required.[83] Large railroads divide their network into hundreds of segments, and determine inspection frequency for each segment based on past inspection results, past history of undetected defects, track tonnage, climate (namely temperature), signaling system (or absence) over segment, whether track carries toxic-byinhalation materials, and whether track is within 500 feet of a structure.[84] They have vehicles to detect defects within the steel of the rail, and others that detect defects in track geometry. From 1980 to 2012, railroads reduced the number of accidents releasing hazmat product per 100,000 hazmat carloads from 14 to 1.[85]

As part of the February 2014 agreement with DOT referenced above, railroads will perform one additional internal-rail inspection each year than required by the FRA on routes over which trains carrying 20 or more tank cars of crude oil travel, and will conduct at least two high-tech track geometry inspections over these routes. Congress may want to look into research on track defect detection technology, including the feasibility of installing equipment on locomotives to achieve near-continuous rail testing.[86]

Shortline Track

It is often the case that a Class I railroad, prior to turning over the operation of a line to a shortline, did not maintain it to the same standards as busier mainlines. Shippers using a shortline often do not require higher-speed track because they ship infrequently or because the commodities they ship are not time-sensitive. Thus, shortline track is frequently maintained at a lower standard than Class I railroads' track. The Lac Mégantic, Quebec, and Aliceville, AL, crude oil derailments occurred on shortline track. Members of Congress have been concerned with preserving shortline rail service, reflected in a federal loan program for track rehabilitation and improvement and a tax credit for shortline track maintenance.[87]

Railroad Operations

A number of specific operational issues have been found relevant to railroad safety, in general, or to oil by rail transportation specifically.

Terminal Operations

In September 2013, the FRA solicited public comment on whether current regulations concerning transfer of crude oil from and to tank cars are adequate considering recent practices at transload facilities. Its request for public comment asked for information about what entity controls trains on loop tracks at rail loading terminals and what procedures have been adopted to prevent unintended movement during loading.[88]

Railroad Crew Size

Following the Lac Mégantic disaster, legislation (H.R. 3040) was introduced in Congress to require two-person crews on all trains. In the United States, the FRA does not specify in regulation how many persons must operate a train, but notes that the various tasks required while a train is moving essentially necessitate at least a two-person crew. Most trains operate with an engineer and a conductor, but some shortline railroads may operate trains with a single crew member. The FRA has announced it intends to issue a proposed rule requiring two-person crews while allowing for some exceptions.[89] One potential trade-off is that distraction by a fellow crew member has been found to be a factor in past accidents.[90]

Positive Train Control

Railroads are in the process of implementing positive train control (PTC), a system that is designed to override human error in controlling the speed and movement of trains. Congress required that this system be installed on routes carrying passengers or poison- or toxic-byinhalation hazardous materials (Section 104 of P.L. 110-432), a requirement that applies to about 60,000 miles of railroad. Current law does not require installation of PTC solely because a track carries crude oil, but the law authorizes the FRA to expand the scope of tracks required to have PTC. PTC is not required on track in or near rail yards. The cost and timeline for implementing PTC are topics of current debate among policy makers and stakeholders.[91]

Route Selection

In the Implementing Recommendations of the 9/11 Commission Act of 2007 (P.L. 110-53, Section 1551), Congress required railroads carrying certain kinds and quantities of potentially dangerous commodities to assess the safest and most secure routes for trains carrying these products and to minimize delays and storage for rail cars containing these products. These requirements currently apply to explosive, toxic-by-inhalation, and radioactive material.[92] Security regulations also require that rail cars containing these commodities not be left unattended when being transferred from one carrier to another or between carrier and shipper.[93] The law resulted from efforts by cities like Washington, DC, and Pittsburgh to ban trains carrying hazardous materials.[94] The FRA may consider whether this routing analysis should also apply to unit trains of crude oil.[95] As part of the February 2014 agreement with DOT mentioned above, railroads agreed to perform this routing analysis for oil trains beginning July 1, 2014. Such a requirement would be controversial because avoiding large urban areas can increase the length of time such trains are in transit and because smaller towns and rural areas likely have less capability to respond to emergencies than large cities. Also, it is unclear to what extent alternative routes are available.

Incident and Oil Spill Response

The increased use of rail for crude oil shipments is likely to increase the number of incidents, some of which may involve oil spills. As described above, the National Oil and Hazardous Substances Contingency Plan provides a framework for federal, state and local collaboration in response to releases of oil and hazardous substances. Considering the relative proximity of rail shipments to population centers, a potential issue for Congress is the safety and adequacy of spill response.

In addition, based on past history, increased frequency or severity of incidents related to shipments of crude oil by rail could lead some local communities to seek additional funding to ensure adequate spill response capabilities, including personnel, training, equipment, and community notification.

The Accuracy of Train Cargo Information

Crude oil may sometimes be carried by "mixed trains"—trains carrying a variety of different commodities. With mixed trains, it is important to first responders that they have an accurate list of which cars contain what commodities (the train "consist"). Often the sequencing of cars changes en route, so the consist information provided by the crew at the scene of an incident may no longer be accurate. Although all vehicles containing hazardous materials must display placards indicating their potential dangerous characteristics (e.g., flammable, corrosive, explosive), responders often need more specific information about the commodities involved in an incident. One potential remedy under consideration is an electronic manifest system that would offer the capability of easier updates. In MAP-21, Congress authorized PHMSA to conduct pilot projects on paperless hazmat information sharing among carriers (of various modes including rail) and first responders.[96] A potential drawback raised by the railroads is that electronic devices at the scene of an incident could encounter technical problems. Another remedy is greater diligence by railroad crew in keeping the paper consist up to date. The NTSB has asked whether a copy of the consist should also be kept at the end of a train in case the copy kept by the crew at the head of the train is lost in an incident.

Rail vs. Pipeline Development

Certain rail routings of crude oil could be replaced by reconfiguring the existing pipeline network and constructing additional pipeline capacity. In general, pipelines could provide safer, less expensive transportation than railroads, assuming that pipeline developers are able to assure markets for the oil they hope to carry.

Pipeline development could be particularly important for shipments of crude oil from Canada to the United States. In light of growing Canadian exports, several proposals have been made to expand the cross-border pipeline infrastructure. Of the five major pipelines currently linking Canadian petroleum producing regions to markets in the United States, two (Alberta Clipper and Keystone) began service in 2010. A permit application for a sixth pipeline, Keystone XL, a very large project which would also transport some Bakken crude, was initially submitted in 2008 and is in the final stages of review by the U.S. Department of State.[97] Keystone XL has been the subject of intense scrutiny and debate by Congress, the executive branch, and numerous stakeholders. The Keystone XL review and approval process is highly contested, and the pipeline's approval remains uncertain.

Other proposed oil pipeline projects, such as the reversal of the Portland-Montreal oil pipeline to enable export of Canadian oil via a marine terminal in Maine, are also encountering greater public scrutiny and opposition. On the whole, the barriers to new oil pipeline approval in any jurisdiction seem to have risen significantly since Alberta Clipper and Keystone were completed.

Shipment of oil by rail is, in many cases, an alternative to new pipeline development. This involves tradeoffs in terms of both transportation capacity and safety. In its ongoing review of the Keystone XL pipeline proposal, the State Department has argued that, if the pipeline is not constructed, additional oil-by-rail capacity will be developed instead. As the State Department's 2014 Final Environmental Impact Statement for the Keystone XL project states,

> In the past 2 years, there has been exponential growth in the use of rail to transport crude oil throughout North America, primarily originating from the Bakken in North Dakota and Montana, but also increasingly utilized in other production areas, including the [Western Canadian Sedimentary Basin]. Because of the flexibility of rail delivery points, once loaded onto trains the crude oil could be delivered to refineries, terminals, and/or port facilities throughout North America, including the Gulf Coast area.[98]

Consistent with this view, both Canadian National Railway and Canadian Pacific Railway reportedly have been pursuing a "pipeline on rails" business strategy, including new track investments, to move Canadian crudes to new markets throughout North America.[99] Increasing cross-border movements of crude oil by rail on existing track does not require State Department approval, so such an approach seeks to avoid regulatory delays. While the potential volumes associated with rail transportation of crude could be lower than pipeline volumes, they could still be significant. Some analysts have suggested that oil-by-rail volumes could be large enough to make a major new pipeline project like Keystone XL unnecessary.[100] Similar arguments could apply to other oil transportation corridors within North America.

Others are less certain that oil by rail can substitute so readily for pipeline capacity, as rail expansion would require significant infrastructure development including loading and unloading facilities, track capacity, and, possibly, additional tank car availability. The State Department's analysis finds that under certain conditions, including particular oil and oil transportation prices, "there could be a substantial impact on oil sands production levels."[101] Other market analyses similarly find that in the short and medium term some production could be curtailed.[102]

Refiner economics may ultimately favor pipelines over rail, although those investment decisions will be determined by market forces. When it comes to safety, however, the federal government plays a major role, and thus may have considerable influence on infrastructure expansion. Some participants in the Keystone XL debate, for example, have asserted that recent oil-by-rail incidents underscore the need for a new pipeline as, in their view, a safer mode of transportation for Canadian crudes,[103] while others insist that safety comparisons between the two transportation modes are less conclusive.[104] On balance, however, it seems likely that policies that raise the cost of transporting oil by rail would increase the attractiveness of pipeline development, and, for that matter, expansion of crude oil transportation by barges, tanker ships, and tanker trucks.

Rail vs. Waterborne Transport

As indicated above, the cost of transporting oil along the coasts in Jones Act tankers is much less than by railroad. However, the fleet of Jones Act-qualified tankers is insufficient to take advantage of this lower-cost shipping method. Despite the domestic oil boom, coastal refineries continue to rely on foreign sources of oil that are shipped at rates generally ranging from less than a dollar per barrel (Mexico to the Gulf Coast) to less than $2.50 a barrel (Saudi Arabia to the Atlantic Coast).[105] In addition to efficient transport, safety is a concern. Tankers are not a new transport method. Tankers are double-hulled, operators are required to have resources and equipment nearby in case of a spill, and the Coast Guard has a regulatory regime in place to promote safe transits through harbors. While the risk of an oil spill in a marine environment remains a grave concern, coastal transport largely avoids travel through towns and cities. The railroads have had difficulty in dealing with the surge in oil traffic, and other rail users are experiencing severe service disruptions in the upper Midwest. An important but open question is whether more oil would be moving by coastwise shipping, relieving some of the safety and capacity pressure on railroads, if tankers were available and their operating costs more competitive. The CEO of Phillips 66 has stated, "I think our view is that because of the limitations on Jones Act vessels, that's going to push you to more barge and more rail, just to evacuate the Gulf Coast."[106]

Rail Transport and Crude Oil Exports

The large increase in U.S. oil production has led some Members of Congress to advocate changing the law that generally prohibits exports of crude oil.[107] An increase in crude oil exports would likely require greater use of rail transportation, as the crude oil pipeline network is not oriented to serve export ports. Some environmental groups have stated their opposition to construction of new rail facilities or terminals that would facilitate oil exports, as they believe increased exports will encourage environmentally damaging production in the United States and Canada.[108]

End Notes

[1] The terms "oil sands" and "tar sands" are often used interchangeably to describe a particular type of nonconventional oil deposit. Opponents of the resource's development often use the term "tar sands," which arguably carries a negative connotation; proponents typically refer to the material as oil sands. The use of this term is not intended to reflect a point of view, but to adopt the term most commonly used by the primary executive-branch agencies involved in recent oil sands policy issues.

[2] Edward R. Hamberger and Andrew J. Black, "Freight Rail and Pipelines Deliver Energy for America," *The Hill, Congress Blog*, November 5, 2013, http://thehill.com/blogs/

[3] *Oilgram Price Report*, "North American Crude By Rail Rising: BNSF," v. 92, no. 58, March 26, 2014, p. 1.

[4] Estimates by CRS based on data from Dagmar Etkin, *Analysis of U.S. Oil Spillage*, API Publication 356, August 2009, and Association of Oil Pipelines, *Report on Shifts in Petroleum Transportation: 1990-2009*, February 2012.

[5] See, for example, Office of Senator John Hoeven, "Hoeven to Meet Saturday with BNSF Railway President and CEO to Address Railroad Safety," press release, January 3, 2014.

[6] Energy Information Administration, *U.S. Crude Oil Supply & Disposition*, http://www.eia.gov/dnav/pet/ pet_sum_crdsnd_k_a.htm. A barrel of oil is equal to 42 gallons.

[7] CRS Report R43128, *Oil Sands and the Oil Spill Liability Trust Fund: The Definition of "Oil" and Related Issues for Congress*, by Jonathan L. Ramseur.

[8] "US Crude Production Tops Imports For The First Time Since 1995," *Oil Daily*, November 14, 2013.

[9] M. Tennyson et al., *Assessment of Remaining Recoverable Oil in Selected Major Oil Fields of the Permian Basin, Texas and New Mexico*, 2012, USGS, http://pubs.usgs.gov/fs/2012/3051/.

[10] "Light" refers to oils with low specific gravity. "Sweet" refers to oils with low sulfur content. Light, sweet crudes are more valuable than heavier or sourer crude oils.

[11] For further information on the petroleum refining industry, refer to CRS Report R41478, *The U.S. Oil Refining Industry: Background in Changing Markets and Fuel Policies*, by Anthony Andrews et al.

[12] Energy Information Administration crude oil production data, by state, available at http://www.eia.doe.gov.

[13] For more information about the Keystone XL pipeline, see CRS Report R41668, *Keystone XL Pipeline Project: Key Issues*, by Paul W. Parfomak et al.

[14] Energy Information Administration price data available at http://www.eia.doe.gov.

[15] Matthew Phillips, "North Dakota's Bakken Oil Finally Hits the East Coast," *Bloomberg Businessweek*, February 6, 2013.

[16] Pipeline data from PHMSA, railroad mileage from Association of American Railroads (includes shortline rail mileage, does not include parallel trackage).

[17] See 49 C.F.R. §213.9.

[18] STB, Docket no. EP 724, April 10, 2004.

[19] BB&T Capital Markets, "Examining The Crude By Barge Opportunity," June 10, 2013, p. 15.

[20] Tank car numbers from presentations by panel on tank car safety (panel 1), NTSB forum, *Rail Safety: Transportation of Crude Oil and Ethanol*, April 22-23, 2014.

[21] E. Russell Braziel, RBN Energy Inc. presentation at CSIS conference, *North American Oil and Gas Infrastructure, Shale Changes Everything*," November 14, 2013.

[22] The bow of the tug fits into a notch in the stern of the barge and the tug is hinged to the barge on both sides of its hull, allowing fore and aft (pitch) movement, such as over sea swells.

[23] Platts, *Oilgram Price Report*, January through April 2014 issues. Railroad distance approximated using Rand McNally Road Atlas.

[24] Platts, *Oilgram News*, September 9, 2013; Bloomberg Businessweek, "U.S. Law Restricting Foreign Ships Leads to Higher Gas Prices," December 12, 2013. The sailing distance is 1,900 nautical miles (one nautical mile equals 1.151 statute miles); National Oceanic and Atmospheric Administration, *Distances Between United States Ports*, 2012.

[25] En*Vantage, Inc., "The Surge in U.S. Crude Oil Production," Presentation to PFAA 20th Annual Conference, October 24, 2013; Bloomberg, "Texas Vies with Saudi Arabian Oil in California Shipments," January 29, 2014. When expansion of the locks through the Panama Canal is completed in 2015, the capacity of tankers able to pass through will increase from 380,000 barrels to 600,000 barrels.

[26] Grain and feed producers in the upper Midwest contend that while they can move product economically by barge to New Orleans or by rail to a Great Lakes port, from there, because of the Jones Act, they have no economic access to dry bulk ships that could deliver the feed

to eastern North Carolina hog and poultry farms. These farms import their feed from Canada and South America.

[27] The law is codified at chapters 81, 121, and 551 of Title 46, *United States Code*.

[28] See for instance, "Oil and the Ghost of 1920," *Wall Street Journal*, September 13, 2012; Senate Committee on Energy and Natural Resources, Testimony of Faisel Khan, Managing Director, Integrated Oil and Gas Research, Citigroup. Hearing to Explore the Effects of Ongoing Changes in Domestic Oil Production, Refining and Distribution on U.S. Gasoline and Fuel Prices, July 16, 2013.

[29] U.S. Maritime Administration, Title XI Ship Financing Guarantees, Pending and Approved Loan Applications; American Petroleum Tankers S-1 SEC Filing; RS Platou Economic Research, annual and monthly reports; press releases from Teekay Tankers, Scorpio Tankers, and Euronav.

[30] U.S. Maritime Administration, U.S. Flag Privately Owned Merchant Fleet, Oceangoing Self-propelled Vessels.

[31] "Phillips 66 Charters Tankers To Bring Shale Oil To Bayway," *Argus Media*, December 13, 2012.

[32] Phillips 66 presentation at Bank of America Merrill Lynch Refining Conference, March 6, 2014.

[33] U.S. Northeast refineries are clustered around New York Harbor and the Delaware River.

[34] This situation is somewhat similar to the Pacific Northwest lumber industry in the 1960s and 1970s, which asserted it could not compete with western Canadian lumber because the Canadians could ship at lower international freight rates to the U.S. East Coast. Washington and Oregon still load significant amounts of wood products on ships, but they all sail to foreign destinations.

[35] Shipping rates from Platts, *Oilgram Price Report*, January through April 2014 issues.

[36] Sources consulted include NOAA, Oil Spill Case Histories, 1967-1991, Summaries of Significant U.S. and International Spills, 1992; U.S. Coast Guard, Notable Spills in U.S. Waters, Calendar Years 1989-2008, 2009; Dagmar Etkin, Analysis of U.S. Oil Spillage, API Publication 356, August 2009; NOAA, Incident News, at http://incidentnews.gov; EPA, Enforcement and Compliance History Online (ECHO), at http://www.epa-echo.gov/echo/index.html.

[37] National Research Council, *Oil in the Sea III: Inputs, Fates, and Effects* (Washington, DC: National Academies of Science, February 2003).

[38] These requirements are codified at 49 CFR §174.85.

[39] Data from Canada's National Energy Board. See also CRS Report R43128, *Oil Sands and the Oil Spill Liability Trust Fund: The Definition of "Oil" and Related Issues for Congress*, by Jonathan L. Ramseur.

[40] The primary vehicle for these arguments was a 2011 report from several environmental groups. See Anthony Swift et al., *Tar Sands Pipelines Safety Risks*, Joint Report by Natural Resources Defense Council, National Wildlife Federation, Pipeline Safety Trust, and Sierra Club, February 2011.

[41] See, e.g., Crude Quality Inc., *Report regarding the U.S. Department of State Supplementary Draft Environmental Impact Statement*, May 2011; and Energy Resources Conservation Board, Press Release, "ERCB Addresses Statements in Natural Resources Defense Council Pipeline Safety Report," February 2011.

[42] P.L. 112-90, §16.

[43] National Research Council, *Effects of Diluted Bitumen on Crude Oil Transmission Pipelines*, 2013.

[44] National Transportation Safety Board, *Accident Report: Enbridge Incorporated Hazardous Liquid Pipeline Rupture and Release- Marshall, Michigan, July 25, 2010*, July 2012, at http://www.ntsb.gov/.

[45] For more up-to-date information, see EPA's Enbridge oil spill website at http://www.epa.gov/ enbridgespill/ index.html.

[46] Letter from Cynthia Giles, Environmental Protection Agency, to U.S. Department of State, April 22, 2013.

[47] EPA Removal Order, March 14, 2013, at http://www.epa.gov/enbridgespill/ar/enbridge-AR-1720.pdf.

[48] See Enbridge Inc., Third Quarter Financial Report, 2013, at http://enbridge.com/ InvestorRelations/FinancialInformation/InvestorDocumentsandFilings.aspx.

[49] Based on cost estimates prepared in 2004. See Dagmar Etkin, *Modeling Oil Spill Response and Damages Costs*, Proceedings of the 5th Biennial Freshwater Spills Symposium, 2004, at http://www.environmental-research.com.

[50] Bryden, K. J., Grace Catalysts Technologies, Columbia, Maryland; Habib Jr., E. T., Grace Catalysts Technologies, Columbia, Maryland; Topete, O. A., Grace Catalysts Technologies, Houston, Texas, Processing shale oils in FCC: Challenges and opportunities 09.01.2013 http://www.hydrocarbonprocessing.com/Article/3250397/Processing-shaleoils-in-FCC-Challenges-and-opportunities.html.

[51] Pipeline and Hazardous Materials Safety Administration, Safety Alert—January 2, 2014, Preliminary Guidance from OPERATION CLASSIFICATION. This advisory is a follow-up to the PHMSA and Federal Railroad Administration (FRA) joint safety advisory published November 20, 2013 [78 FR 69745].

[52] FRA Administrator Szabo, Opening Remarks to RSAC Meeting, October. 31, 2013; http://www.fra.dot.gov/eLib/ Details/L04852.

[53] FRA and PHMSA are agencies within DOT, which has the emergency authority to restrict or prohibit transportation that poses a hazard of death, personal injury, or significant harm to the environment. See 49 U.S.C. §20104.

[54] The NTSB held a forum on the safety of crude oil and ethanol transport by rail April 22 and 23, 2014; http://www.ntsb.gov/news/events/2014/railsafetyforum/.

[55] 40 C.F.R. Part 112.

[56] 40 C.F.R. §300.5. This definition is also found in the Clean Water Act and OPA.

[57] The Oil Pollution Act of 1990 expanded and clarified the President's authorities under Section 311 of the Clean Water Act (33 U.S.C. §2701 et. seq.). For a more in-depth discussion of the Oil Pollution Act, see CRS Report RL33705, *Oil Spills in U.S. Coastal Waters: Background and Governance*, by Jonathan L. Ramseur.

[58] 33 U.S.C. §1321. In addition, the Comprehensive Environmental Response, Compensation, and Liability Act (CERCLA) of 1980 expanded the authorities of the President to respond to releases of hazardous substances into the environment more broadly than CWA Section 311. See CRS Report R41039, *Comprehensive Environmental Response, Compensation, and Liability Act: A Summary of Superfund Cleanup Authorities and Related Provisions of the Act*, by David M. Bearden. For further details, see CRS Report R43251, *Oil and Chemical Spills: Federal Emergency Response Framework*, by David M. Bearden and Jonathan L. Ramseur.

[59] 33 U.S.C. §1321(c).

[60] Executive Order 12777, "Implementation of Section 311 of the Federal Water Pollution Control Act of October 18, 1972, as amended, and the Oil Pollution Act of 1990," 56 *Federal Register* 54757, October 22, 1991.

[61] 49 C.F.R. Part 130.

[62] See 49 C.F.R. §130.31(a) with 49 C.F.R. §130.31(b).

[63] 61 *Federal Register* 30541 (June 17, 1996).

[64] NTSB, Safety Recommendation R-14-4 through -6, directed to PHMSA, January 21, 2014.

[65] U.S. Congress, House Committee on Transportation and Infrastructure, Subcommittee on Railroads, Pipelines, and Hazardous Materials, *Oversight of Passenger and Freight Rail Safety*, 113th Cong., 2nd sess., February 26, 2014; U.S. Congress, Senate Committee on Commerce, Science, and Transportation, Subcommittee on Surface Transportation and Merchant Marine Infrastructure, Safety, and Security, *Enhancing Our Rail Safety: Current Challenges for Passenger and Freight Rail*, 113th Cong., 2nd sess., March 6, 2014; U.S. Congress, Senate Committee on Appropriations, Subcommittee on Transportation and Housing and Urban Development, and Related Agencies, *Rail Safety*, 113th Cong., 2nd sess., April 9, 2014.

[66] See FRA's Emergency Order No. 28 (78 *Federal Register* 48218), the agencies' Joint Safety Advisory published August 7, 2013 (78 *Federal Register* 48224), referral of safety issues to FRA's Railroad Safety Advisory Committee (78 *Federal Register* 48931), and a NPRM related to rail hazmat (78 *Federal Register* 54849).

[67] Xiang Liu, M. Rapik Saat, Christopher P.L. Barkan, *Safety Effectiveness of Integrated Risk Reduction Strategies for the Transportation of Hazardous Materials by Rail*, paper presented at the Transportation Research Board, Annual Meeting 2013, paper no. 13-1811.

[68] AAR, "Freight Railroads Join U.S. Transportation Secretary Foxx in Announcing Industry Crude By Rail Safety Initiative," February 21, 2014.

[69] Athaphon Kawprasert and Christopher P.L. Barkan, "Effect of Train Speed on Risk Analysis of Transporting Hazardous Materials by Rail," *Transportation Research Record*, No. 2159, 2010, pp. 59-68.

[70] The tank cars used to transport crude oil fall under DOT specification 111. See 49 C.F.R. §179.

[71] 49 C.F.R. §179.201.

[72] For a discussion of NTSB's recommendations concerning DOT-111 tank cars, in reference to the derailment of an ethanol unit train in Cherry Valley, IL, see NTSB Safety Recommendation R-12-5 through -8, March 2, 2012.

[73] Pressurized tank cars (DOT specification 105 and 112) have thicker shells and heads, metal jackets, strong protective housings for top fittings, and no bottom valves.

[74] 78 FR 54849 - 54861, September 6, 2013.

[75] For comments filed on this rulemaking see http://www.regulations and search under docket no. PHMSA-2012- 0082.

[76] See letter from Senator Schumer to PHMSA and FRA dated July 22, 2013 and news release by Senator Hoeven on January 15, 2013 indicating that a DOT final rule on tank cars would not be issued until after January 2015.

[77] Transport Canada, "Addressing the safety of DOT-111 tank cars carrying dangerous goods," April 23, 2014.

[78] CQ News, "Oil Train Regulations Couldn't Come Soon Enough For Some Railroads," May 1, 2014.

[79] T87.6 Task Force Summary Report, pp. 9-11; Xiang Liu, M. Rapik Saat, Christopher P.L. Barkan, "Analysis of Causes of Major Train Derailment and Their Effect on Accident Rates," *Transportation Research Record*, No. 2289, 2012, pp. 154-163.

[80] Xiang Liu, M. Rapik Saat, Christopher P.L. Barkan, *Safety Effectiveness of Integrated Risk Reduction Strategies for the Transportation of Hazardous Materials by Rail*, Paper presented at the Transportation Research Board, Annual Meeting 2013, paper no. 13-1811.

[81] 79 *Federal Register* 4234, January 24, 2014.

[82] This risk element has been studied by Xiang Liu, University of Illinois at Urbana-Champaign—RailTEC, Presentation at Transportation Research Board Annual Meeting, Session 279, *Broken Rail Prevention and Rail Flaw Assessment*, Washington, DC, January 13, 2014.

[83] 79 *Federal Register* 4234 and 4245, January 24, 2014.

[84] Presentations of BNSF and UP Railroads, Transportation Research Board Annual Meeting, Washington, DC, January 13, 2014, Session 279, *Broken Rail Prevention and Rail Flaw Assessment.*

[85] Christopher P.L. Barkan, M. Rapik Saat, and Francisco Gonzalez III et al., "Cooperative Research in Tank Car Safety Design," *TR News*, vol. 286 (May-June 2013), pp. 12-19.

[86] This topic was briefly discussed at the NTSB forum on rail safety cited above.

[87] The Railroad Rehabilitation and Improvement Financing (RRIF) program and Section 45G of the tax code.

[88] "FRA/PHMSA Additional Questions for Public Comment," Docket No. FRA-2013-0067-0016, 9/4/2013, http://www.regulations.

[89] Press Release no. FRA-03-14, April 9, 2014.

[90] NTSB, *Collision of Two CN Freight Trains, Anding, Mississippi, July 10, 2005*, Accident Report RAR-07/01, p. 31.

[91] For further information, see CRS Report R42637, *Positive Train Control (PTC): Overview and Policy Issues*, by John Frittelli.

[92] See 49 C.F.R. §172.820; 73 *Federal Register* 72182, November 26, 2008.

[93] See 49 C.F.R. §1580.107.

[94] *U.S. Rail News*, June 11, 2008, pp. 1-2; "Hazmat Hazards: U.S. Cities may not wait for Washington Before Trying to Reroute their own hazmat trains," *Journal of Commerce,* December 12, 2005.

[95] RSAC meeting, presentation by HAZMAT Working Group, October 31, 2013. The NTSB has recommended this change; see Safety Recommendation R-14-1 through -3, January 23, 2014.

[96] Section 33005 of P.L. 112-141.

[97] The construction, connection, operation, and maintenance of a pipeline connecting the United States with a foreign country require executive permission through a Presidential Permit under Executive Orders 11423 and 13337.

[98] U.S. Department of State, January 2014, Final EIS, Section 5.1, "No Action Alternatives."

[99] Nathan Vanderklippe, "CN, CP Push for a 'Pipeline on Rails,'" *Globe and Mail*, February 7, 2011.

[100] "Keystone Pipeline Seen as Unneeded as More Oil Moves by Rail," *CBC News*, September 10, 2013.

[101] 2014 Final EIS, p. 1.4-8.

[102] For example, Canadian Imperial Bank of Commerce, "Too Much of A Good Thing: A Deep Dive Into The North American Energy Renaissance," August 15, 2012; TD Economics, "Pipeline Expansion is a National Priority," Special Report, December 17, 2012; International Energy Agency," Medium-Term Oil Market Report," May 14, 2013.

[103] Diana Furchtgott-Roth and Kenneth P. Green, *Intermodal Safety in the Transport of Oil*, Fraser Institute, October 2013, http://www.fraserinstitute.org/uploadedFiles/fraser-ca/Content/research-news/research/publications/intermodalsafety-in-the-transport

[104] See, for example: Rory Johnston, "Train vs. Pipeline: What's the Safest Way to Transport Oil?" *Christian Science Monitor*, Energy Voices blog, October 22, 2013, http://www.

csmonitor.com/Environment/Energy-Voices/2013/1022/ Train-vs.-pipeline-What-s-the-safest-way-to-transport-oil.

[105] Platts, *Oilgram Price Report*, January through April 2014.

[106] Phillips 66 Earnings Conference Call, Q2 2013, July 31, 2013.

[107] The Senate Energy and Natural Resources Committee held a hearing on this issue on January 30, 2014. For further information and analysis on oil exports, see CRS Report R43442, *U.S. Crude Oil Export Policy: Background and Considerations*, by Phillip Brown et al.

[108] See, for example, the comments of Sierra Club official Michael Marx in Blake Sobczak, "Environmentalists 'get real creative' to combat oil by rail," *Energy Wire*, January 13, 2014.

In: Oil and Gas Transportation
Editor: Elton Simmons

ISBN: 978-1-63463-841-8

Chapter 3

KEEPING AMERICA'S PIPELINES SAFE AND SECURE: KEY ISSUES FOR CONGRESS*

Paul W. Parfomak

SUMMARY

Nearly half a million miles of pipeline transporting natural gas, oil, and other hazardous liquids crisscross the United States. While an efficient and fundamentally safe means of transport, many pipelines carry materials with the potential to cause public injury and environmental damage. The nation's pipeline networks are also widespread and vulnerable to accidents and terrorist attack. Recent pipeline accidents in Marshall, MI, San Bruno, CA, Allentown, PA, and Laurel, MT, have heightened congressional concern about pipeline risks and drawn criticism from the National Transportation Safety Board (NTSB). Both government and industry have taken numerous steps to improve pipeline safety and security over the last 10 years. Nonetheless, while many stakeholders agree that federal pipeline safety programs have been on the right track, the spate of recent pipeline incidents suggest there continues to be significant room for improvement. Likewise, the threat of terrorist

* This is an edited, reformatted and augmented version of a Congressional Research Service publication R41536, prepared for Members and Committees of Congress, dated January 9, 2013.

attacks, especially cyberattacks on pipeline control systems, remains a concern.

The federal pipeline safety program is authorized through the fiscal year ending September 30, 2015, under the Pipeline Safety, Regulatory Certainty, and Job Creation Act of 2011 (P.L. 112- 90), which was signed by President Obama on January 3, 2012. The act contains a broad range of provisions addressing pipeline safety and security. Among the most significant are provisions that could increase the number of federal pipeline safety inspectors, require automatic shutoff valves for transmission pipelines, mandate verification of maximum allowable operating pressure for gas transmission pipelines, increase civil penalties for pipeline safety violations, and mandate reviews of diluted bitumen pipeline regulation.

Both government and industry have taken numerous steps to improve pipeline safety and security over the last 10 years. Nonetheless, the NTSB has identified improvement of federal pipeline safety oversight as a "top ten" priority for 2013. The leading pipeline industry associations have concurred. Whether renewed efforts by industry, combined with additional oversight by federal agencies, will further enhance the safety and security of U.S. pipelines remains to be seen.

As Congress oversees the federal pipeline safety program and the federal role in pipeline security, key issues of focus may be pipeline agency staff resources, automatic pipeline shutoff valves, penalties for safety violations, safety regulations for oil sands crudes, and the possible need for pipeline security regulations, among other concerns. In addition to these specific issues, Congress may assess how the various elements of U.S. pipeline safety and security activity fit together in the nation's overall strategy to protect transportation infrastructure. Pipeline safety and security necessarily involve many groups: federal agencies, oil and gas pipeline associations, large and small pipeline operators, and local communities. Reviewing how these groups work together to achieve common goals could be an oversight challenge for Congress.

INTRODUCTION

Nearly half a million miles of high-volume pipeline transport natural gas, oil, and other hazardous liquids across the United States.[1] These transmission pipelines are integral to U.S. energy supply and have vital links to other critical infrastructure, such as power plants, airports, and military bases. While an efficient and fundamentally safe means of transport, many pipelines carry volatile, flammable, or toxic materials with the potential to cause public injury and environmental damage. The nation's pipeline networks are also widespread, running alternately through remote and densely populated regions, some above ground and some below. These systems are vulnerable to accidents and terrorist attack. Recent pipeline accidents in Marshall, MI, San Bruno, CA, Allentown, PA, and Laurel, MT, have demonstrated this vulnerability and have heightened congressional concern about pipeline risks.

The federal program for pipeline safety resides primarily within the Department of Transportation (DOT), although its inspection and enforcement activities rely heavily upon partnerships with state pipeline safety agencies. The federal pipeline security program began with the DOT as well, immediately after the terror attacks of September 11, 2001, but pipeline security authority was subsequently transferred to the Department of Homeland Security (DHS) when the latter department was created. The DOT and DHS have distinct missions, but they cooperate to protect the nation's pipelines. The Federal Energy Regulatory Commission is not operationally involved in pipeline safety or security, but it can examine safety issues under its siting authority for interstate natural gas pipelines, and can allow pipeline companies under its rate jurisdiction to recover pipeline security costs. Collectively, these agencies administer a comprehensive and complex set of regulatory authorities which has been changing significantly over the last decade and continues to do so.

The federal pipeline safety program is authorized through the fiscal year ending September 30, 2015, under the Pipeline Safety, Regulatory Certainty, and Job Creation Act of 2011 (P.L. 112- 90), which was signed by President Obama on January 3, 2012. This report reviews the history of federal programs for pipeline safety and security, key policy issues, and recent developments leading up to P.L. 112-90. Although the act contains over 30 substantive sections, this report focuses only on a subset of provisions of key interest in recent congressional debate.

SAFETY AND SECURITY IN THE PIPELINE INDUSTRY

Of the nation's approximately half million miles of transmission pipeline, roughly 180,000 miles carry hazardous liquids—over 75% of the nation's crude oil and around 60% of its refined petroleum products, along with other products.[2] Within this network, there are nearly 200 *inter*state crude oil and liquid fuel pipelines, which account for roughly 80% of total pipeline mileage and transported volume.[3]

The U.S. natural gas pipeline network consists of around 300,000 miles of *inter*state and *intra*state transmission.[4] It also contains some 14,000 miles of onshore field and gathering pipeline, which connect gas extraction wells to processing facilities.[5] Around 120 systems make up the *inter*state gas transmission network; another 90 or so systems operate strictly within individual states.[6] These *inter*state and *intra*state gas transmission pipelines feed around 900,000 miles of regional pipelines in some 1,400 local distribution networks serving over 66 million customers.[7] Natural gas pipelines also connect to 127 active liquefied natural gas (LNG) storage sites, which can augment pipeline gas supplies during peak demand periods.[8]

Pipeline Safety Record

Taken as a whole, releases from pipelines cause few annual fatalities compared to other product transportation modes.[9] According to the DOT, there were 14 deaths per year on average from all U.S. pipeline systems from 2007 through 2011.[10] Accidental pipeline releases result from a variety of causes, including third-party excavation, corrosion, mechanical failure, control system failure, and operator error. Natural forces, such as floods and earthquakes, can also damage pipelines. There were 140 hazardous liquid pipeline accidents, 84 natural gas transmission (including gathering) pipeline accidents, and 58 natural gas distribution accidents in 2011.[11]

Although pipeline releases have caused relatively few fatalities in absolute numbers, a single pipeline accident can be catastrophic in terms of deaths and environmental damage. Notable pipeline accidents in recent years include:

- *1999*—A gasoline pipeline explosion in Bellingham, WA, killed three people and caused $45 million in damage to a city water plant and other property.
- *2000*—A natural gas pipeline explosion near Carlsbad, NM, killed 12 campers.
- *2006*—Corroded pipelines on the North Slope of Alaska leaked over 200,000 gallons of crude oil in an environmentally sensitive area and temporarily shut down Prudhoe Bay oil production.
- *2007*—An accidental release from a propane pipeline and subsequent fire near Carmichael, MS, killed two people, injured several others, destroyed four homes, and burned over 70 acres of grassland and woodland.

- *2010*—A pipeline spill in Marshall, MI, released 819,000 gallons of crude oil into a tributary of the Kalamazoo River.
- *2010*—A natural gas pipeline explosion in San Bruno, CA, killed 8 people, injured 60 others, and destroyed 37 homes.
- *2011*—A natural gas pipeline explosion in Allentown, PA, killed 5 people, damaged 50 buildings, and caused 500 people to be evacuated.
- *2011*—A pipeline spill near Laurel, MT, released an estimated 42,000 gallons of crude oil into the Yellowstone River.
- *2012*—A natural gas pipeline explosion in Springfield, MA injured 21 people and heavily damaged over a dozen buildings.

Such accidents have generated persistent scrutiny of pipeline regulation and have increased state and community activity related to pipeline safety.

Pipeline Security Vulnerabilities

In addition to their vulnerability to accidents, pipelines may also be intentionally damaged by vandals and terrorists. Pipelines may also be vulnerable to "cyber-attacks" on supervisory control and data acquisition (SCADA) systems or attacks on electricity grids and communications networks.[12] Oil and gas pipelines, globally, have been a favored target of terrorists, militant groups, and organized crime. In Colombia, for example, rebels have bombed the Caño Limón oil pipeline and other pipelines over 950 times since 1993.[13] In 1996, London police foiled a plot by the Irish Republican Army to bomb gas pipelines and other utilities across the city.[14] Militants in Nigeria have repeatedly attacked pipelines and related facilities, including the simultaneous bombing of three oil pipelines in May 2007.[15] A Mexican rebel group similarly detonated bombs along Mexican oil and natural gas pipelines in July and September 2007.[16] In June 2007, the U.S. Department of Justice arrested members of a terrorist group planning to attack jet fuel pipelines and storage tanks at the John F. Kennedy (JFK) International Airport in New York.[17] Natural gas pipelines in British Columbia, Canada, were bombed six times between October 2008 and July 2009 by unknown perpetrators.[18] In 2009, the *Washington Post* reported that over $1 billion of crude oil had been stolen directly from Mexican pipelines by organized criminals and drug cartels.[19]

Since September 11, 2001, federal warnings about Al Qaeda have mentioned pipelines specifically as potential terror targets in the United States.[20] One U.S. pipeline of particular concern, and with a history of terrorist and vandal activity, is the Trans Alaska Pipeline System (TAPS), which transports crude oil from Alaska's North Slope oil fields to the marine terminal in Valdez. TAPS runs some 800 miles and delivered approximately 600,000 barrels of oil per day in 2011 (over 10% of U.S. domestic oil production).[21] In 1999, Vancouver police arrested a man planning to blow up TAPS for personal profit in oil futures.[22] In 2001, a vandal's attack on TAPS with a high-powered rifle forced a two-day shutdown and caused extensive economic and ecological damage.[23] In January 2006, federal authorities acknowledged the discovery of a detailed posting on a website purportedly linked to Al Qaeda that reportedly encouraged attacks on U.S. pipelines, especially TAPS, using weapons or hidden explosives.[24] In November 2007 a U.S. citizen was convicted of trying to conspire with Al Qaeda to attack TAPS and a major natural gas pipeline in the eastern United States.[25]

Notwithstanding the incidents cited above, to date, there have been no known Al Qaeda attacks on TAPS or other U.S. pipelines. The most recent U.S. federal threat assessment concludes "with high confidence that the terrorist threat to the U.S. pipeline industry is low ... [with] no specific or credible threat information indicating that violent transnational extremist groups or domestic extremists are actively plotting to conduct attacks on the U.S. pipeline industry."[26] Terrorist activities are in constant flux, however, and difficult to predict, so terrorist attacks remain a possibility in the future. Although Al Qaeda attacks on U.S. pipelines may be perceived as unlikely, attacks by individuals not associated with organized or terrorist groups may be a growing concern. For example, in August 2011, federal agents arrested a U.S. citizen—acting alone—who confessed to planting an improvised explosive device (which failed to detonate) under a natural gas pipeline in Oklahoma.[27] In June 2012, a man was critically injured attempting to plant an improvised explosive device along a natural gas pipeline in Plano, TX.[28]

Cybersecurity Risks[29]

One specific area of pipeline security risk that has recently come to the fore is SCADA system cybersecurity. In March 2012, the Industrial Control Systems Cyber Emergency Response Team within DHS identified an ongoing series of cyber intrusions among U.S. natural gas pipeline operators dating back to December 2011. According to the agency, various pipeline companies described targeted spear-phishing[30] attempts and intrusions into multiple natural gas pipeline sector organizations "positively identified ... as related to a single campaign."[31] In 2011, computer security company McAfee reported similar "coordinated covert and targeted" cyberattacks originating primarily in

China against global energy companies. The attacks began in 2009 and involved spear-phishing, exploitation of Microsoft software vulnerabilities, and the use of remote administration tools to collect sensitive competitive information about oil and gas fields.[32] In 2010, the Stuxnet computer worm was first identified as a threat to industrial control systems. Although the Stuxnet software initially spreads indiscriminately, the software includes a highly specialized industrial process component targeting specific Siemens industrial SCADA systems.[33] Computer security specialists claim that malicious software developers have already created new software programs tailored to target the kinds of SCADA system weaknesses revealed by Stuxnet.[34] The increased vulnerability of pipeline SCADA systems due to their modernization, taken together with the emergence of SCADA-specific malicious software and the recent cyberattacks, suggests that cybersecurity threats to pipelines have been increasing.

Pipelines and Hazardous Materials Safety Administration

The Natural Gas Pipeline Safety Act of 1968 (P.L. 90-481) and the Hazardous Liquid Pipeline Act of 1979 (P.L. 96-129) are two of the principal early acts establishing the federal role in pipeline safety. Under both statutes, the Transportation Secretary is given primary authority to regulate key aspects of interstate pipeline safety: design, construction, operation and maintenance, and spill response planning. Pipeline safety regulations are covered in Title 49 of the *Code of Federal Regulations*.[35] The DOT administers pipeline regulations through the Office of Pipeline Safety (OPS) within the Pipelines and Hazardous Materials Safety Administration (PHMSA). At the end of FY2012, PHMSA employed 203 total staff, including 135 inspection and enforcement staff.[36] In addition to its own staff, PHMSA's enabling legislation allows the agency to delegate authority to *intra*state pipeline safety offices, and allows state offices to act as "agents" administering *inter*state pipeline safety programs (excluding enforcement) for those sections of *inter*state pipelines within their boundaries.[37] Approximately 350 state pipeline safety inspectors were available in 2012.[38]

PHMSA's pipeline safety program is funded primarily by user fees assessed on a per-mile basis on each regulated pipeline operator.[39] P.L. 109-468 authorized annual pipeline safety program expenditures of $79.0 million in FY2007, $86.2 million in FY2008, $91.5 million in FY2009, and $96.5 million in FY2010. P.L. 112-90 authorizes expenditures of $109.3 million annually for each of FY2012 through FY2015, and $1.5 million annually through FY2015 for state pipeline damage prevention programs. The President's FY2013 budget requested $177 million.[40]

PHMSA uses a variety of strategies to promote compliance with its safety standards. The agency conducts programmatic inspections of management systems, procedures, and processes; conducts physical inspections of facilities and construction projects; investigates safety incidents; and maintains a dialogue with pipeline operators. The agency clarifies its regulatory expectations through published protocols and regulatory orders, guidance manuals, and public meetings. PHMSA relies upon a range of enforcement actions, including administrative actions such as corrective action orders (CAOs) and civil penalties, to ensure that operators correct safety violations and take measures to preclude future safety problems. Between January 1 and November 8, 2012, PHMSA initiated 244 enforcement actions against pipeline operators.[41] Civil penalties proposed by PHMSA for safety violations during this period totaled approximately $8.3 million.[42] PHMSA also conducts accident investigations and system-wide reviews focusing on high-risk operational or procedural problems and areas of the pipeline near sensitive environmental areas, high-density populations, or navigable waters.

Since 1997, PHMSA has increasingly required industry's implementation of "integrity management" programs on pipeline segments near "high consequence areas." Integrity management provides for continual evaluation of pipeline condition; assessment of risks to the pipeline; inspection or testing; data analysis; and follow-up repair; as well as preventive or mitigative actions. High consequence areas include population centers, commercially navigable waters, and environmentally sensitive areas, such as drinking water supplies or ecological reserves. The integrity management approach directs priority resources to locations of highest consequence rather than applying uniform treatment to the entire pipeline network. PHMSA made integrity management programs mandatory for most oil pipeline operators with 500 or more miles of regulated pipeline as of March 31, 2001 (49 C.F.R. §195).

DOT Pipeline Security Activities

Presidential Decision Directive 63 (PDD-63), issued during the Clinton Administration, assigned lead responsibility for pipeline security to the DOT.[43] These responsibilities fell to the OPS, at that time part of the DOT's Research and Special Programs Administration (RSPA), since the agency was already addressing some elements of pipeline security in its role as safety regulator.[44] In 2002, the OPS conducted a vulnerability assessment to identify critical pipeline facilities and worked with industry groups and state pipeline safety organizations "to assess the industry's readiness to prepare for, withstand and respond to a terrorist attack."[45] Together with the Department of Energy and state pipeline agencies, the OPS promoted the development of consensus standards for security measures tiered to correspond with the five levels of threat warnings issued by the Office of Homeland Security.[46] The OPS also developed protocols for inspections of critical facilities to ensure that operators implemented appropriate security practices. To convey emergency information and warnings, the OPS established communication links to key staff at the most critical pipeline facilities throughout the country. The OPS also began identifying near-term technology to enhance deterrence, detection, response, and recovery, and began seeking to advance public and private sector planning for response and recovery.[47]

On September 5, 2002, the OPS circulated guidance developed in cooperation with the pipeline industry defining the agency's security program recommendations and implementation expectations. This guidance recommended that operators identify critical facilities, develop security plans consistent with prior trade association security guidance, implement these plans, and review them annually.[48] Although the guidance was voluntary, the OPS expected compliance and informed operators of its intent to begin reviewing security programs within 12 months, potentially as part of more comprehensive safety inspections.[49] Federal pipeline security authority was subsequently transferred outside of DOT, however, as discussed below, so the OPS did not follow through on a national program of pipeline security program reviews.

Pipeline Safety Improvement Act of 2002

On December 12, 2002, President Bush signed into law the Pipeline Safety Improvement Act of 2002 (P.L. 107-355). The act strengthened federal pipeline safety programs, state oversight of pipeline operators, and public education regarding pipeline safety.[50] Among other provisions, P.L. 107-355 required operators of regulated natural gas pipelines in high-consequence areas to conduct risk analysis and implement integrity management programs similar to those required for oil pipelines.[51] The act authorized the DOT to order safety actions for pipelines with potential safety problems and increased violation penalties. The act streamlined the permitting process for emergency pipeline restoration by establishing an interagency committee, including the DOT, the Environmental Protection Agency, the Bureau of Land Management, the Federal Energy Regulatory Commission, and other agencies, to ensure coordinated review and permitting of pipeline repairs. The act required DOT to study ways to limit pipeline safety risks from population encroachment and ways to preserve environmental resources in pipeline rights-of-way. P.L. 107-355 also included provisions for public education, grants for community pipeline safety studies, "whistle blower" and other employee protection, employee qualification programs, and mapping data submission.

Pipeline Inspection, Protection, Enforcement, and Safety Act of 2006

On December 29, 2006, President Bush signed into law the Pipeline Inspection, Protection, Enforcement and Safety Act of 2006 (PIPES Act, P.L. 109-468). The main provisions of the act address pipeline damage prevention, integrity management, corrosion control, and enforcement transparency. The PIPES act created a national focus on pipeline damage prevention through grants to states for improving damage prevention programs, establishing 811 as the national "call before you dig" one-call telephone number, and giving PHMSA limited "backstop" authority to conduct civil enforcement against one-call violators in states that have failed to conduct such enforcement. The act mandated the promulgation by PHMSA of minimum standards for integrity management programs for natural gas distribution pipelines.[52] It also mandated a review of the adequacy of federal pipeline safety regulations related to internal corrosion control, and required PHMSA to increase the transparency of enforcement actions by issuing monthly summaries, including violation and penalty information, and a mechanism for pipeline operators to make response information available to the public.

Transportation Security Administration

In November 2001, President Bush signed the Aviation and Transportation Security Act (P.L. 107- 71) establishing the Transportation Security Administration (TSA) within the DOT. According to TSA, the act placed the DOT's pipeline security authority (under PDD-63) within TSA. The act specified for TSA a range of duties and powers related to general transportation security, such as intelligence management, threat assessment, mitigation, security measure oversight and enforcement, among others. On November 25, 2002, President Bush signed the Homeland Security Act of 2002 (P.L. 107-296) creating the Department of Homeland Security (DHS). Among other provisions, the act transferred to DHS the Transportation Security Administration from the DOT (§403). On December 17, 2003,

President Bush issued Homeland Security Presidential Directive 7 (HSPD-7), clarifying executive agency responsibilities for identifying, prioritizing, and protecting critical infrastructure.[53] HSPD-7 maintains DHS as the lead agency for pipeline security (par. 15), and instructs the DOT to "collaborate in regulating the transportation of hazardous materials by all modes (including pipelines)" (par. 22h). The order requires that DHS and other federal agencies collaborate with "appropriate private sector entities" in sharing information and protecting critical infrastructure (par. 25). TSA joined both the Energy Government Coordinating Council and the Transportation Government Coordinating Council under provisions in HSPD-7. The missions of the councils are to work with their industry counterparts to coordinate critical infrastructure protection programs in the energy and transportation sectors, respectively, and to facilitate the sharing of security information.

HSPD-7 also required DHS to develop a national plan for critical infrastructure and key resources protection (par. 27), which the agency issued in 2006 as the *National Infrastructure Protection Plan* (NIPP). The NIPP, in turn, required each critical infrastructure sector to develop a Sector Specific Plan (SSP) that describes strategies to protect its critical infrastructure, outlines a coordinated approach to strengthen its security efforts, and determines appropriate funding for these activities. Executive Order 13416 further required the transportation sector SSP to prepare annexes for each mode of surface transportation.[54] In accordance with the above requirements the TSA issued its *Transportation Systems Sector Specific Plan* and *Pipeline Modal Annex* in 2007 with an update on 2010.

TSA Pipeline Security Activities

Pipeline security activities at TSA are led by the Pipeline Security Division (PSD) within the agency's Office of Security Policy and Industry Engagement (OSPIE).[55] According to the agency's *Pipeline Modal Annex* (PMA), TSA has been engaged in a number of specific pipeline security initiatives as summarized in *Table 1*.

In addition to the activities in *Table 1*, TSA has also conducted regional supply studies for key natural gas markets, has conducted training on cyber security awareness, has participated in pipeline blast mitigation studies, and has joined in "G-8" multinational security assessment and planning.[56]

In 2003, TSA initiated its Corporate Security Review (CSR) program, wherein the agency visits the largest pipeline and natural gas distribution operators to review their security plans and inspect their facilities.

Table 1. Ongoing TSA Pipeline Security Initiatives

Initiative	Description	Participants
Pipeline System Relative Risk	Statistical tool used for relative risk ranking and prioritizing	TSA, Industry
Tool	CSR findings	
Corporate Security Reviews	On-site reviews of pipeline operator security	TSA, Industry
(CSR)		
Critical Facility Inspections	On-site inspection of critical facilities of the 100 most	TSA, Industry
(CFI)	critical pipeline systems	
Pipeline Security Guidance	Maintains and revises voluntary pipeline security guidelines	TSA, Other agencies,
	and best practices with input from government and industry partners	Industry
Security Incident and	Completed a security incident and recovery protocol plan	TSA, Other agencies,
Recovery	mandated under P.L. 110-53	Industry
Toxic Inhalation Hazard Materials Program	Program to address potential risks from pipeline transportation of hazardous materials other than oil and natural gas	TSA, Other agencies, Industry
Pipeline Cross-Border Vulnerability Assessment	U.S. and Canadian security assessment and planning for critical cross-border pipeline	TSA, Canada
International Pipeline Security Forums	International forums for U.S. and Canadian governments and pipeline industry officials convened annually	TSA, Canada, Other agencies, Industry
Pipeline Security Exercises	Facilitation of pipeline security drills and exercises including those under the Intermodal Security Training Exercise Program (I-STEP)	TSA, Other Agencies, Industry
Security Awareness Training Materials	Informational compact discs about pipeline security issues and improvised explosive devices	TSA
Stakeholder Conference Calls	Periodic information-sharing conference calls between key pipeline security stakeholders	TSA, Other agencies, Industry
Sector Coordinating Councils and Joint Sector Committee	Government partners coordinate interagency and cross-jurisdictional implementation of critical infrastructure security	TSA, Other agencies, Industry

Sources: Transportation Security Administration, Transportation Systems Sector-Specific Plan, 2010, pp. 326.

During the reviews, TSA evaluates whether each company is following the intent of the OPS security guidance, and seeks to collect the list of assets each company had identified meeting the criteria established for critical facilities. In 2004, the DOT reported that the plans reviewed to date (approximately 25) had been "judged responsive to the OPS guidance."[57] TSA has completed CSRs covering the largest 100 pipeline systems (84% of total U.S. energy pipeline throughput) and is in the process of conducting second CSRs of these systems.[58] According to TSA, CSR results indicate that the majority of U.S. pipeline systems "do a good job with pipeline security" although there are areas in which pipeline security can be improved.[59] Past CSR reviews have identified inadequacies in some company security programs such as not updating security plans, lack of management support, poor employee involvement, inadequate threat intelligence, and employee apathy or error.[60] In 2008, the TSA initiated its Critical Facility Inspection Program (CFI), under which the agency conducted in-depth inspections of all the critical facilities of the 125 largest pipeline systems in the United States. The agency estimated that these 125 pipeline systems collectively included approximately 600 distinct critical facilities.[61] TSA concluded the CFI program in May 2011, having completed a total of 347 inspections throughout the United States.[62]

In addition to the initiatives in *Table 1*, TSA has worked to establish qualifications for personnel applying for positions with unrestricted access to critical pipeline assets and has developed its own inventory of critical pipeline infrastructure.[63] The agency has also addressed legal issues regarding recovery from terrorist attacks, such as FBI control of crime scenes and eminent domain in pipeline restoration. In October 2005, TSA issued an overview of recommended security practices for pipeline operators "for informational purposes only ... not intended to replace security measures already implemented by individual companies."[64] The agency released revised pipeline security guidelines in 2010 and 2011. The guidelines include a section on cybersecurity developed with the assistance of the Applied Physics Laboratory of Johns Hopkins University as well as other government and industry stakeholders.[65]

The President's FY2012 budget request for DHS did not include a separate line item for TSA's pipeline security activities. The budget request did include a $134.7 million line item for "Surface Transportation Security," which encompasses security activities in non-aviation transportation modes, including pipelines.[66] The PSD has traditionally received from the agency's general operational budget an allocation for routine operations such as regulation development, travel, and outreach. According to the PSD, the budget funds 13 full-time equivalent staff within the office.[67]

In 2007 the TSA Administrator testified before Congress that the agency intended to conduct a pipeline infrastructure study to identify the "highest risk" pipeline assets, building upon such a list developed through the CSR program. He also stated that the agency would use its ongoing security review process to determine the future implementation of baseline risk standards against which to set measurable pipeline risk reduction targets.[68] Provisions in the Implementing Recommendations of the 9/11 Commission Act of 2007 (P.L. 110-53) require TSA, in consultation with PHMSA, to develop a plan for the federal government to provide increased security support to the "most critical" pipelines at high or severe security alert levels and when there is specific security threat information relating to such pipeline infrastructure (§1558(a)(1)).

The act also requires a recovery protocol plan in the event of an incident affecting the interstate and intrastate pipeline system (§1558(a)(2)). TSA published this plan in 2010.

Security Incident Investigations

In addition to the above pipeline security initiatives, the TSA Pipeline Security Division has performed a limited number of vulnerability assessments and has supported investigations for specific companies and assets where intelligence information has suggested potential terrorist activity. The PSD, along with PHMSA, was involved in the investigation of an August 2006 security breach at an LNG peak-shaving plant in Lynn, MA.[69] Although not a terrorist incident, the security breach involved the penetration of intruders through several security barriers and alert systems, permitting them to access the main LNG storage tank at the facility. The PSD also became aware of the JFK airport terrorist plot in its early stages and supported the Federal Bureau of Investigation's associated investigation. The PSD engaged the private sector in helping to assess potential targets and determine potential consequences. The PSD worked with the pipeline company to keep it informed about the plot, discuss its security practices, and review its emergency response plans.[70]

GAO Study of TSA' s Pipeline Security Activities

In December 2008, the Senate Committee on Commerce, Science, and Transportation requested a study by the Government Accountability Office (GAO) examining TSA's efforts to ensure pipeline security. GAO's report, released in August 2010, focused on TSA's use of risk assessment and risk information in securing pipelines, actions the agency has taken to improve pipeline security under guidance in the 9/11 Commission Act of 2007 (P.L. 110-53), and the agency's efforts to measure such security improvement efforts.[71] Among other findings, GAO concluded that, although TSA had begun to implement a risk management approach to prioritize its pipeline security efforts, work remained to ensure that the highest risk pipeline systems would get the necessary scrutiny. GAO also concluded that TSA was missing opportunities under its CSR and CFI programs to better ensure that pipeline operators understand how they can enhance the security of their pipeline systems. TSA could also make better use of CSR and CFI recommendations for analyzing pipeline vulnerabilities and was not following up on these recommendations. GAO found that linking TSA's pipeline security performance measures and milestones to the goals and objectives in its national security strategy for pipeline systems could aid in achieving results within specific time frames and could facilitate more effective oversight and accountability.[72] TSA concurred with all of GAO's recommendations for addressing the issues and has since been implementing them.[73]

Federal Energy Regulatory Commission

One area related to pipeline safety and security not under either PHMSA's or TSA's primary jurisdiction is the siting approval of new gas pipelines, which is the responsibility of the Federal Energy Regulatory Commission (FERC). Companies building interstate natural gas pipelines must first obtain from FERC certificates of public convenience and necessity. (FERC does not oversee oil pipeline construction.) FERC must also approve the abandonment of gas facility use and services. These approvals may include safety and security provisions with respect to pipeline routing, safety standards, and other factors.[74] As a practical matter, however, FERC has traditionally left these considerations to the other agencies.[75]

On September 14, 2001, FERC notified jurisdictional companies that it would "approve applications proposing the recovery of prudently incurred costs necessary to further safeguard the nation's energy systems and infrastructure" in response to the terror attacks of 9/11. FERC also committed to "expedite the processing on a priority basis of any application that would specifically recover such costs from wholesale customers." Companies could propose a surcharge over currently existing rates or some other cost recovery method.[76] In FY2005, the commission processed security cost recovery requests from 14 oil pipelines and 3 natural gas pipelines.[77] FERC's FY2006 annual report stated that "the Commission continues to give the highest priority to deciding any requests made for the recovery of extraordinary expenditures to safeguard the reliability and security of the Nation's energy transportation systems and energy supply infrastructure."[78] FERC's subsequent annual reports do not mention pipeline security.

In February 2003, FERC promulgated a new rule (RM02-4-000) to protect critical energy infrastructure information (CEII). The rule defines CEII as information that "must relate to critical infrastructure, be potentially useful to terrorists, and be exempt from disclosure under the Freedom of Information Act." According to the rule, critical infrastructure is "existing and proposed systems and assets, whether physical or virtual, the incapacity or destruction of which would negatively affect security, economic security, public health or safety, or any combination of those matters." CEII excludes "information that identifies the location of infrastructure." The rule also establishes procedures for the public to request and obtain such critical information, and applies both to proposed and existing infrastructure.[79]

On May 14, 2003, FERC handed down new rules (RM03-4) facilitating the restoration of pipelines after a terrorist attack. The rules allow owners of a damaged pipeline to use blanket certificate authority to immediately start rebuilding, regardless of project cost, even outside existing rights-of-way. Pipeline owners would still need to notify landowners and comply with environmental laws. Prior rules limited blanket authority to projects up to $17.5 million and 45- day advance notice.[80]

National Transportation Safety Board

The National Transportation Safety Board (NTSB) is an independent federal agency charged with determining the probable cause of transportation accidents (including pipeline accidents), promoting transportation safety, and assisting accident victims and their families. The board's experts investigate significant accidents, develop factual records, and issue safety recommendations to prevent similar accidents from recurring. The NTSB has no statutory authority to regulate transportation, however, so its safety recommendations to industry or government agencies are not mandatory. Nonetheless, because of the board's strong reputation for thoroughness and objectivity, the average acceptance rate for its safety recommendations is

78%.[81] The NTSB's "Most Wanted List" for 2013 called for enhanced pipeline safety through improved oversight of the pipeline industry.[82]

San Bruno Pipeline Accident Investigation

In August 2011, the NTSB issued preliminary findings and recommendations from its investigation of the San Bruno Pipeline accident. The investigation included testimony from pipeline company officials, government agency officials (PHMSA, state, and local), as well as testimony from other pipeline experts and stakeholders. In addition to specifics about the San Bruno incident, the hearing addressed more general pipeline issues, including public awareness initiatives, pipeline technology, and oversight of pipeline safety by federal and state regulators.[83] The NTSB's findings were highly critical of the pipeline operator (PG&E) as well as both the state and federal pipeline safety regulators. The board concluded that "the multiple and recurring deficiencies in PG&E operational practices indicate a systemic problem" with respect to its pipeline safety program.[84] The board further concluded that

> the pipeline safety regulator within the state of California, failed to detect the inadequacies in PG&E's integrity management program and that the Pipeline and Hazardous Materials Safety Administration integrity management inspection protocols need improvement. Because the Pipeline and Hazardous Materials Safety Administration has not incorporated the use of effective and meaningful metrics as part of its guidance for performance-based management pipeline safety programs, its oversight of state public utility commissions regulating gas transmission and hazardous liquid pipelines could be improved.

In her opening statement about the San Bruno accident report, NTSB Chairman Hersman summarized the board's findings as "troubling revelations ... about a company that exploited weaknesses in a lax system of oversight and government agencies that placed a blind trust in operators to the detriment of public safety."[85] The NTSB's final accident report "concludes that PHMSA's enforcement program and its monitoring of state oversight programs have been weak and have resulted in the lack of effective Federal oversight and state oversight."[86]

The NTSB issued 39 recommendations stemming from its San Bruno accident investigation, including 20 recommendations to the Secretary of Transportation and PHMSA. These recommendations included:

- Conducting audits to assess the effectiveness of PHMSA's oversight of performance-based pipeline safety programs and state pipeline safety program certification,
- Requiring pipeline operators to provide system-specific information to the emergency response agencies of the communities in which pipelines are located,
- Requiring that automatic shutoff valves or remote control valves be installed in high consequence areas and in class 3 and 4 locations,[87]
- Requiring that all natural gas transmission pipelines constructed before 1970 be subjected to a hydrostatic pressure test that incorporates a spike test,
- Requiring that all natural gas transmission pipelines be configured so as to accommodate internal inspection tools, with priority given to older pipelines, and
- Revising PHMSA's integrity management protocol to incorporate meaningful metrics, set performance goals for pipeline operators, and require operators to regularly assess the effectiveness of their programs using meaningful metrics.[88]

More detailed discussion of the accident findings and the NTSB's recommendations are publicly available in the final accident report.

KEY POLICY ISSUES

The 112th Congress reauthorized the federal pipeline safety program and enacted a number of new pipeline safety provisions. In the context of its continuing oversight of federal pipeline safety and security activities, and in addition to the findings of the NTSB's San Bruno investigation, the 113th Congress may focus on certain key issues that have drawn particular attention in recent policy deliberations.

Staffing Resources for Pipeline Safety and Security

The U.S. pipeline safety program employs a combination of federal and state staff to implement and enforce federal pipeline safety regulations. To date, PHMSA has relied heavily on state agencies for pipeline inspections, with only approximately 25% of inspectors in 2012 being federal employees. Some in Congress have criticized inspector staffing at PHMSA as being insufficient to adequately cover pipelines under the agency's jurisdiction, notwithstanding state agency cooperation. In considering PHMSA staff levels, three distinct issues are the overall number of federal inspectors, the agency's historical use of staff funding, and the staffing of pipeline safety inspectors among the states.

PHMSA Inspectors

The President's FY2013 budget request listed PHMSA's estimated staffing in 2013 as 290 full-time equivalent employees (FTEs), up from 206 funded staff in 2010. By contrast, as *Figure 1* shows, the addition of 10 inspection and enforcement staff under P.L. 112-90 amounts to only a modest continuation of staff growth (of mostly inspectors) begun over 10 years ago in response to the 1999 Bellingham accident, the terrorist attacks of 9/11, implementation of PHMSA's integrity management regulations, and the expansion of U.S. pipeline networks.

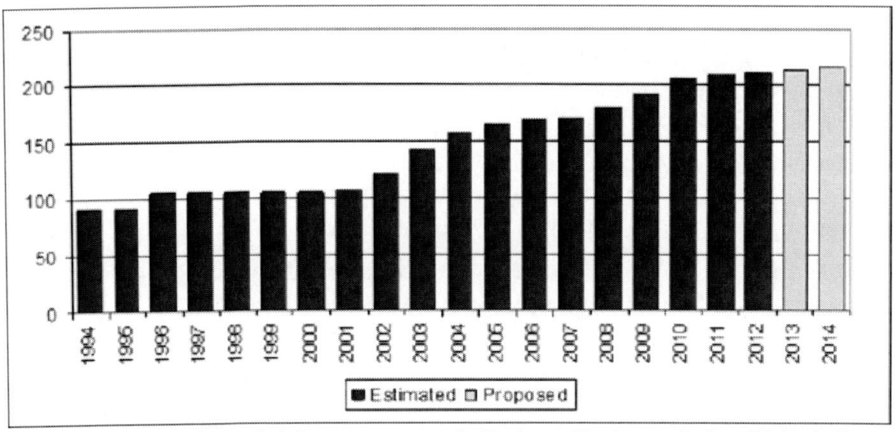

Sources: U.S. Office of Management and Budget, Budget of the United States Government: Appendix, Fiscal Years1996-2011; P.L. 112-90 §31(b).

Notes: Estimated staff are staff anticipated by the agency as reported in annual budget requests. They may differ from actual staff employed (for the same fiscal year) as reported in subsequent budget requests.

Figure 1. PHMSA Pipeline Safety Staffing, Historical and Proposed under P.L. 112-90. (Full-Time Equivalent Staff)

Whether an increase of 10 PHMSA pipeline safety staff by 2014, in addition to filling all previously authorized positions, would be the optimal number remains to be seen. Under the President's FY2013 budget request for 290 FTEs, most of the staff additions would presumably be inspectors.[89] While such an increase would represent a dramatic increase in the number of federal pipeline safety staff available, filling all these positions and retaining all these employees might pose practical challenges, further discussed below.

PHMSA Staffing Shortfalls

One issue that has complicated the PHMSA staffing debate is a long-term pattern of understaffing in the agency's pipeline safety program. At least as far back as 1994, PHMSA's (or RSPA's) actual staffing for pipeline safety as reported in its annual budgets requests has generally fallen well short of the level of staffing anticipated in the prior year's budget request. For example, the president's FY2011 budget request for pipeline safety reports 175 actual employees in 2009. However, the FY2010 budget request stated an expectation of 191 employees ("estimated") for 2009. On this basis, from 2001 through 2011, the agency reported a staffing shortfall averaging approximately 23 employees every year. (Note that, due to this annual shortfall, the FTEs reported in *Figure 1* are generally higher than the number actually employed by PHMSA.) Furthermore, most of this staffing shortage has been among inspectors. In testimony before Congress in September 2010, DOT officials reported that PHMSA employed only 110 of 137 inspectors for which it was funded—a shortfall of 27 inspectors.[90] In March 2011, agency officials reported 126 inspectors employed.[91] However, as of the end of FY2012, PHMSA reported 135 inspection and enforcement staff out of 203 total staff at the agency—nearly a full complement of funded inspector FTE's.[92]

PHMSA officials offer a number of reasons for the historical shortfall in inspector staffing. These reasons include a scarcity of qualified inspector job applicants, delays in the federal hiring process during which applicants accept other job offers, and PHMSA inspector turnover— especially to pipeline companies which often hire away PHMSA inspectors for their corporate safety programs. Because PHMSA pipeline inspectors are highly trained by the agency (typically for two years before being allowed to operate independently) they are highly valued by pipeline operators seeking to comply with federal safety regulations. The agency states that it aggressively recruits a qualified and diverse workforce but is challenged by industry recruitment of the same candidates, especially with the rapid development of unconventional oil and gas shales, for which the skill sets PHMSA seeks (primarily engineers) are in high demand.[93] PHMSA officials also cite structural issues associated with the agency's appropriations, which can require the use of FTE salary funding to meet other obligations.[94] P.L. 112-90 requires the DOT to report on PHMSA's pipeline staffing shortfalls, including the reasons for such shortfalls, and actions the agency is taking to fill the positions (§31(a)). This report has not yet been completed.

Table 2. Actual vs. Anticipated Pipeline Safety Staff in DOT Budget Requests. (Full-Time Equivalent Staff)

Year	Actual	Anticipated	Difference
1994	62	90	-28
1995	71	90	-19
1996	84	105	-21
1997	92	105	-13
1998	97	105	-8
1999	93	105	-12
2000	97	105	-8
2001	96	107	-11
2002	100	122	-22
2003	111	143	-32
2004	125	156	-31
2005	154	164	-10
2006	139	169	-30
2007	146	170	-24
2008	147	180	-33
2009	162	191	-16
2010	175	206	-31
2011	193	206	-13
2012	203	215	-12

Sources: U.S. Office of Management and Budget, Budget of the United States Government: Appendix, Fiscal Years1996-2013; Linda Daugherty, Pipeline and Hazardous Materials Safety Administration, personal communication, December 13, 2012; CRS analysis.

State Pipeline Inspector Funding

Because state agencies would continue to account for the majority of U.S. pipeline safety inspectors under P.L. 112-90, another important consideration is how the number of state inspectors might be affected by budget shortfalls and possible agency funding cuts faced by many states due to the recent U.S. economic recession. Under P.L. 109-468 (§2(c)), PHMSA is authorized to award grants reimbursing state governments for up to 80% of the cost of the staff, personnel, and activities required to support the federal pipeline safety program. According to DOT, these grant are essential to "enable the states to continue their current programs and hire additional inspectors ... [and] assure that states do not turn over responsibility for distribution pipeline systems to the Federal inspectors," among other reasons.[95] According to PHMSA, in 2012, the average state grant was approximately 74% of state program expenditures during 2011.[96]

Notwithstanding federal pipeline safety grants, inspector staffing at state pipeline safety agencies has been negatively affected by state budget deficits. According to a 2010 letter from the National Association of Pipeline Safety Representatives, pipeline safety employees in 17 states had already been furloughed without pay for up to three weeks at that time.[97] PHMSA officials have also reportedly cited unfilled positions among state pipeline safety agencies as eroding the state pipeline safety workforce.[98] Senior DOT officials consider financial problems among state pipeline safety agencies a matter of "great concern" and have granted to states waivers from certain regulatory financial requirements to increase their access to federal grant money.[99] Nonetheless, the future availability of state pipeline safety inspectors remains uncertain. In particular, the possibility that some states may choose to end their roles as agents for the federal pipeline safety program—or that states may lose federal pipeline safety program certification for performance reasons—and thereby shift a greater burden for pipeline inspections back to the federal government, may warrant continued attention from Congress.

TSA Pipelines Security Resources

Similar to its concerns about the adequacy of federal pipeline safety staffing, Congress has long been concerned about staff resources available to implement the nation's pipeline security program. For example, as one Member remarked in 2005, "aviation security has received 90% of TSA's funds and virtually all of its attention. There is simply not enough being done to address ... pipeline security."[100] At a congressional field hearing in April 2010, another Member expressed concern that TSA's pipeline division did not have sufficient staff to carry out a federal pipeline security program on a national scale.[101]

At its current staffing level of 13 FTEs, TSA's Pipelines Security Division has limited field presence. In conducting a pipeline corporate security review, for example, TSA typically sends one to three staff to hold a three- to four-hour interview with the operator's security representatives, followed by a visit to only one or two of the operator's pipeline assets.[102] TSA's plan to focus security inspections on the largest pipeline and distribution system operators tries to make the best use of its limited resources. However, there are questions as to whether the agency's CSRs as currently structured allow for rigorous security plan verification and a credible threat of enforcement. The limited number of CSRs the agency can complete in a year is a particular concern. According to a 2009 GAO report, "TSA's pipeline division stated that they would like more staff in order to conduct its corporate security reviews more frequently," and "analyzing secondary or indirect consequences of a terrorist attack and developing strategic risk objectives required much time and effort."[103]

Since both PHMSA and TSA have played important roles in the federal pipeline security program, with TSA the designated lead agency since 2002, Congress has raised questions about the appropriate responsibilities and division of pipeline security authority between them.[104] According to TSA, the two agencies have established "a 24/7 communication and coordination relationship in regards to all pipeline security and safety incidents."[105] Nonetheless, given the limited staff in TSA's pipeline security division, and the comparatively large pipeline safety staff (especially inspectors) in PHMSA, legislators have considered whether the TSA-PHMSA pipeline security relationship optimally aligns staff resources across both agencies to fulfill the nation's overall pipeline safety and security mission.[106] The Transportation Security Administration Authorization Act of 2011 (H.R. 3011) in the 112th Congress would have mandated a study regarding the relative roles and responsibilities of the Department of Homeland Security and the Department of Transportation with respect to pipeline security (§325).[107]

Automatic Shutoff Valves for Transmission Pipelines

In the 2010 San Bruno pipeline accident, natural gas continued to flow from the pipeline for nearly two hours after the initial explosion—fueling the intense fire, hindering emergency response, and increasing fire damage. The long duration of flowing gas was due to delays in the closing of manually operated valves by the pipeline operator, and may have been exacerbated by inadequate employee training in valve closure procedures.[108] Consequently, some advocates have called for widespread installation of remotely or automatically controlled valves in natural gas and hazardous liquids transmission pipelines. As noted earlier, the NTSB has recommended the installation of such valves in all "high consequence" and relatively more populated areas. P.L. 112-90 requires automatic or remote-controlled shut-off valves, or equivalent technology, where economically, technically, and operationally feasible on transmission pipeline facilities constructed or entirely replaced (§4(1)). The act also requires a study on the ability of transmission pipeline operators to respond to a hazardous liquid or natural gas release from pipelines located in high-consequence areas (§4(2)).

Previous Consideration

The possibility of requiring remotely controlled or automatic shutoff valves for natural gas pipelines is not new. Congress previously considered such requirements in reaction to a 1994 natural gas pipeline fire in Edison, NJ, similar to the San Bruno accident, in which it took the pipeline operator 21/2 hours to close its manually operated valves.[109] In 1995, during the 104th Congress, H.R. 432 and S. 162 would have required the installation of remotely or automatically controlled valves in natural gas pipelines "wherever technically and economically feasible" (§11). Under the Accountable Pipeline Safety and Partnership Act of 1996 (P.L. 104-304), Congress mandated a DOT assessment of remotely controlled valves (RCVs) on interstate natural gas pipelines, and empowered the agency to require such valves if appropriate based upon its findings (§4(h)).

The DOT's assessment, released in 1999, reported that installation of RCVs would provide only "a small benefit from reduced casualties because virtually all casualties from a rupture occur before an RVC could be activated."[110] Moreover, the DOT reported that it lacked data to compare pipeline fire property damage with and without RCVs. Nonetheless, the DOT study advocated the deployment of RCVs, at least in some gas pipeline locations.

> We have found that RCVs are effective and technically feasible, and can reduce risk, but are not economically feasible. We have also found that there may be a public perception that RCVs will improve safety and reduce the risk from a ruptured gas pipeline. We believe there is a role for RCVs in reducing the risk from certain ruptured pipelines and thereby minimizing the consequences of certain gas pipeline ruptures.... Any fire would be of greater intensity and would have greater potential for damaging surrounding infrastructure if it is constantly replenished with gas. The degree of disruption in heavily populated and commercial areas would be in direct proportion to the duration of the fire. Although we lack data enabling us to quantify

> these potential consequences, we believe them to be significant nonetheless, and we believe RCVs may provide the best means for addressing them.[111]

Notwithstanding this conclusion, the DOT has not mandated the use of RCVs in natural gas transmission pipelines.

The natural gas pipeline industry historically has objected to federal mandates to install remotely controlled or automated valves. Although pipeline operators already employ such valves under specific circumstances, such as in hard-to-access locations or at compressor stations, they have opposed the installation of such valves more widely throughout their pipeline systems on the grounds that they are usually not cost-effective. They also argue that such valves do not always function properly, would not prevent natural gas pipeline explosions (which cause most fatalities), and are susceptible to false alarms, needlessly shutting down pipelines and disrupting critical fuel supplies.[112] Automatic valves, in particular, may be susceptible to unnecessary closure, potentially disrupting critical flows of natural gas to distribution utilities and—as a result—increasing safety risks associated with residential furnace relighting, among other concerns.[113] Some operators also claim higher maintenance costs for valves that are not manually operated.

Remotely Controlled Valves for Liquids Pipelines

The use of remotely controlled or automatic valves has also been a long-standing consideration for hazardous liquid pipeline systems. The National Transportation Safety Board (NTSB) began to address the need for rapid shutdown of failed hazardous liquid pipelines using remotely controlled or automatic valves in the 1970s.[114] In 1987, the NTSB recommended that the DOT "require the installation of remote-operated valves on pipelines that transport hazardous liquids, and base the spacing of remote-operated valves on the population at risk."[115] The Pipeline Safety Act of 1992 (P.L. 102-508) required the DOT to assess the effectiveness of "emergency flow restricting devices (including remotely controlled valves and check valves)" on hazardous liquid pipelines, and required the DOT to "issue regulations prescribing the circumstances under which operators of hazardous liquid pipeline facilities must use emergency flow restricting devices" (§212). Notwithstanding this congressional mandate, the NTSB found the DOT's efforts to promote the use of such devices inadequate. In 1996, the NTSB stated that the DOT "has performed studies, conducted research, and sought industry input, but has failed to carry through and develop requirements for leak detection and rapid shutdown of failed pipelines."[116] In its integrity management regulations, issued in December 2000, the DOT opted to leave the decision whether to install emergency flow restricting devices up to pipeline operators.[117]

Valve Replacement Costs

Cost would be a major factor in a broad national program to retrofit manual valves with remotely controlled or automatic valves. For example, in the interstate natural gas pipeline network, valves are typically installed every 5 to 20 miles. Assuming a 10-mile separation between valves, the nation's 306,000 mile gas transmission system contains over 30,000 valves. The spacing of valves can be much closer together in particular pipeline systems, however, such as systems located in more populated areas. In October 2010 PG&E reported 300 valves that could be candidates for automation in approximately 565 miles of high-consequence area pipelines in its California system.[118]

The potential costs of retrofitting manual valves vary greatly by pipeline and specific location. A 1998 Southwest Research Institute report estimated a cost of $32,000 (approximately $40,000 in 2010 dollars) per valve for retrofitting 30-inch pipeline valves to make them remotely controlled.[119] The DOT's 1999 study reported an average cost of $83,000 (approximately $100,000 in 2010 dollars) for Texas Eastern Transmission Corporation (TETCO) to retrofit 90 existing valves in a large part of its pipeline system.[120] PG&E estimates the average cost of retrofitting an automatic or remotely controlled valve on an existing large diameter pipeline at approximately $750,000, but ranging as low as $100,000 and as high as $1.5 million.[121]

Applying, for illustration, a $100,000 cost to some 30,000 valves yields $3.0 billion in capital investment required, not counting any higher future maintenance expenses. The American Gas Association reportedly has estimated the cost of replacing manual valves with automatic valves nationwide at $12 billion.[122] Even if such valve retrofits were required only in heavily populated areas, industry costs could still be hundreds of millions of dollars—a significant cost to the pipeline industry and therefore likely to increase rates for pipeline transportation of natural gas. To the extent that some pipeline systems, like PG&E's, contain more valves then others per mile of pipe, they could be disproportionately affected. Gas pipeline service interruptions would also be an issue as specific lines could be repeatedly taken out of service during the valve retrofit process. The hazardous liquids pipeline industry could face capital costs and service interruptions of the same magnitude if required to do a widespread valve retrofit on existing lines. Additional right-of-way costs, environmental impacts, and construction accidents associated with the valve replacements could also be a consideration. For new pipelines, the incremental costs of installing remotely controlled or automatic valves instead of manual valves would be lower than in the retrofit case, but could still increase future pipeline costs.

Oak Ridge Laboratory Valve Study

In October 2012, Oak Ridge National Laboratory released a safety study of remotely controlled and automatic pipeline shutoff valves which the laboratory completed for the DOT in response to the requirements in P.L. 112-90 and related NTSB recommendations. Among other findings, the study concluded that such valves can be effective for mitigating potential consequences resulting from a natural gas pipeline release (and subsequent fire). However, because natural gas pipeline fires can cause damage so quickly, such mitigation requires that the leak is detected and the proper valves closed completely so the damaged pipeline segment can be isolated and firefighting activities can begin *within 10 minutes* of the initial fire. Fire hydrants must also be accessible in the vicinity of the leak within the potentially severe fire damage radius.[123] For hazardous liquid pipelines, the study similarly concluded that installing such valves can be effective for mitigating potential fire damage resulting from a "guillotine-type" break and subsequent fire if the leak is detected and the damaged pipeline segment isolated *within 15 minutes* after the break. The report further concluded that "adding automatic closure capability to block valves in newly constructed or fully replaced hazardous liquid pipelines can also be an effective strategy for mitigating potential socioeconomic and environmental damage resulting from a release that does not ignite."[124] Thus the Oak Ridge study concludes that while remotely controlled and automatic pipeline shutoff valves can improve safety, they can do so only in conjunction with rapid and well-coordinated emergency response.

SCADA and Leak Detection System Requirements

To effectively reduce the impact of pipeline accidents, installing remotely controlled or automatic valves may require associated investments in supervisory control and data acquisition (SCADA) systems along with other operational changes to improve leak detection. As one pipeline expert has stated,

> The pipeline operator's focus on keeping the pipeline system operating and the lack of remotely-operable valves are the primary factors that control the quantity of product released after a rupture or leak. Even with remote control valves this relationship will not change unless the pipeline is equipped with a reliable leak detection subsystem that works with the SCADA system and [unless] those who control pipeline operations are trained for and dedicated to minimizing product release (safety and environmental mindset) rather than trained for and dedicated to keeping the system operating (economic mindset).[125]

In its report about a 1996 pipeline accident in Tiger Pass, LA, the NTSB similarly concluded that the operator's "delay in recognition ... that it had experienced a pipeline rupture at Tiger Pass was due to the piping system's dynamics during the rupture and to the design of the company's SCADA system."[126] Estimates of converting manual valves may, therefore, need to account for the costs of SCADA changes, leak detection systems, and associated training. These costs may also include significant reliability and security components, since increasing reliance upon new or expanded SCADA systems may also expose pipeline systems to greater risk from operating software failure or cyberterrorism.[127]

Consistent with the concerns above, P.L. 112-90 requires a DOT analysis of the technical limitations of leak detection systems as well as the feasibility of establishing standards for such systems (§8(a)). After congressional review of this analysis, the act authorizes the DOT to issue new leak detection standards if "practicable" (§8(b)). The agency released its leak detection study in December 2012. However, the study does not reach specific conclusions or recommendations; it only provides data and reports on technical and cost aspects of leak detection systems.[128] The DOT has not stated whether or how it intends to develop new leak detection standards based on the findings of this report.

Public Perceptions

Some stakeholders have argued that public perceptions of improved pipeline safety and control are the highest perceived benefit of remotely controlled or automatic valves.[129] Although the value of these perceptions is hard to quantify (and, therefore, not typically reflected in cost-effectiveness studies), the importance of public perception and community acceptance of pipeline infrastructure can be a significant consideration in pipeline design, expansion, and regulation. In 2001, a representative of the National

Association of Regulatory Utility Commissioners testified before Congress that "the main impediment to siting energy infrastructure is the great difficulty getting public acceptance for needed facilities."[130] Likewise, the National Commission on Energy Policy stated in its 2006 report that energy-facility siting is "a major cross-cutting challenge for U.S. energy policy," largely because of public opposition to new energy projects and other major infrastructure.[131]

One result of public concern about pipeline safety has been to prevent new pipeline siting in certain localities, and to increase pipeline development time and costs in others. In a 2006 report, for example, the Energy Information Administration (EIA) stated that "several major projects in the Northeast, although approved by FERC, have been held up because of public opposition or non-FERC regulatory interventions."[132] In the specific case of the Millennium Pipeline, proposed in 1997 to transport Canadian natural gas to metropolitan New York, developers did not receive final construction approval for nine years, largely because of community resistance to the pipeline route.[133] Numerous other proposed pipelines, especially in populated areas, have faced similar public acceptance barriers.[134] Controversy surrounding the proposed Keystone XL pipeline project, discussed below, is only the most recent example of how the development of major pipeline projects may be influenced by public opinion. Even where there is federal siting authority, as is the case for interstate natural gas pipelines, community stakeholders retain many statutory and regulatory avenues to affect energy infrastructure decisions. Consequently, the public perception value of remotely controlled or automatic pipeline valves may need to be accounted for, especially with respect to its implications for general pipeline development and operations.

Natural Gas Distribution Excess Flow Valves

While the San Bruno, CA, and Edison, NJ, gas pipeline accidents focused attention on automatic valves in large diameter transmission pipelines, this technology also applies to smaller gas distribution lines serving individual buildings. In natural gas distribution systems, "excess flow" valves are safety devices which can automatically shut off pipeline flow in the event of a leak. In this way, the valves can minimize the release of natural gas during a pipeline accident, thereby reducing the likelihood or severity of a fire or explosion. PHMSA issued new standards requiring the installation of excess flow valves on new gas distribution lines in single-family homes as part of its final rule for natural gas distribution integrity management programs on

December 3, 2009.[135] P.L. 112-90 authorizes regulation, "if appropriate," requiring excess flow valves for new or entirely replaced distribution branch pipelines, as well as for service lines to multi-family residential buildings and small commercial facilities (§22). Although smaller in scale, automatic valves in distribution lines raise the same cost and safety tradeoffs as automatic valves in large diameter pipelines.

PHMSA Penalties and Pipeline Safety Enforcement

The adequacy of the PHMSA's enforcement strategy has been an ongoing focus of congressional oversight.[136] Provisions in the Pipeline Safety Improvement Act of 2002 (P.L. 107-355) put added scrutiny on the effectiveness of the agency's enforcement strategy and assessment of civil penalties (§8). In April 2006, PHMSA officials testified before Congress that the agency had institutionalized a "tough-but-fair" approach to enforcement, "imposing and collecting larger penalties, while guiding pipeline operators to enhance higher performance."[137] According to the agency, $4.6 million in proposed civil penalties in 2005 was three times greater than penalties proposed in 2003, the first year higher penalties could be imposed under P.L. 107-355 (§8(a)).[138] Proposed penalties totaled $4.5 million in 2010.[139] Proposed penalties in 2011 totaled $3.7 million, with an average penalty of approximately $65,500.[140] P.L. 112-90 increases the maximum civil penalty from $1.0 million to $2.0 million for a related series of major consequence violations, such as those causing serious injuries, deaths, or environmental harm (§2(a)).

Although PHMSA's imposition of pipeline safety penalties increased quickly after P.L. 107-355 was enacted, the role of federal penalties in promoting greater operator compliance with pipeline safety regulations is not always clear. To understand the potential influence of penalties on operators, it can be helpful to put PHMSA fines in the context of the overall costs to operators of a pipeline release. Pipeline companies, seeking to generate financial returns for their owners, are motivated to operate their pipelines safely (and securely) for a range of financial reasons. While these financial considerations certainly include possible PHMSA penalties, the costs of a pipeline accident may also include fines for violations of environmental laws (federal and state), the costs of spill response and remediation, penalties from civil litigation, the value of lost product, costs for pipeline repairs and modifications (e.g., to resolve federal regulatory interventions), and other costs. Depending upon the severity of a pipeline release, these other costs may far exceed pipeline safety fines, as illustrated by the following examples.

- *Kinder Morgan.* In April 2006 Kinder Morgan Energy Partners entered into a consent agreement with PHMSA to resolve a corrective action order stemming from three hazardous liquid spills in 2004 and 2005 from the company's Pacific Operations pipeline unit.[141] According to the company, the agreement would require Kinder Morgan to spend approximately $26 million on additional integrity management activities, among other requirements.[142] Under a 2007 settlement agreement with the U.S. Justice Department and the State of California, Kinder Morgan also agreed to pay approximately $3.8 million in civil penalties for violations of environmental laws and approximately $1.5 million related to response and remediation associated with these spills. The spills collectively released approximately 200,000 gallons of diesel fuel, jet fuel, and gasoline.[143] This volume of fuel would have a product value on the order of $0.5 million based on typical wholesale market prices at the time of the spills.
- *Plains All American.* In 2010, Plains All American Pipeline agreed to spend approximately $41 million to upgrade 10,420 miles of U.S. oil pipeline to resolve Clean Water Act (CWA) violations for 10 crude oil spills in Texas, Louisiana, Oklahoma, and Kansas from 2004 through 2007. Among these upgrades, the company agreed to spend at least $6 million on equipment and materials for internal corrosion control and surveys on at least 2,400 miles of pipeline. The company was required to pay a $3.25 million civil penalty associated with the CWA violations.[144]
- *Enbridge.* Enbridge Energy Partners estimated expenses of $475 million to clean up two oil spills on its Lakehead pipeline system in 2010, including the spill in Marshall, MI. This estimate did not include fines or penalties which might also be imposed in connection with the spills. The pipeline operator also reported $16 million in lost revenue from pipeline shipments it could not redirect to other lines while the Lakehead system was out of service.[145] The full impact of these expenditures on the company's business is unclear, however, as Enbridge stated in a subsequent quarterly report that "substantially all of the costs" related to its 2010 oil pipeline spills "will ultimately be recoverable under our existing insurance policies."[146]

- *Olympic Pipe Line.* After the 1999 Bellingham pipeline accident, Olympic Pipe Line Company and associated defendants reportedly agreed to pay a $75 million settlement to the families of two children killed in the accident.[147]
- *El Paso.* In 2002, El Paso Corporation settled wrongful death and personal injury lawsuits stemming from the 2000 natural gas pipeline explosion near Carlsbad, NM, which killed 12 campers.[148] Although the terms of those settlements were not disclosed, two additional lawsuits sought a total of $171 million in damages.[149] However, El Paso's June 2003 quarterly financial report stated that "our costs and legal exposure ... will be fully covered by insurance."[150]

PHMSA Penalties in Perspective

The threat of safety enforcement penalties is often considered one of the primary tools available to pipeline safety regulators to ensure operator compliance with safety requirements. However, as the examples above suggest, pipeline safety fines, even if they were raised to $2.0 million for major violations, could still account for only a limited share of the financial impact of future pipeline releases. So, it is not clear how large an effect increasing PHMSA's authorized fines, alone, might have on operator compliance. On the other hand, the authority of PHMSA to influence pipeline operations directly—for example, through corrective action orders or shutdown orders in the event of a pipeline failure—can have a large financial impact on a pipeline operator in terms of capital expenditures or lost revenues. Indeed, some have suggested that this operational authority is the most influential component of PHMSA's pipeline safety enforcement strategy. Therefore, as Congress continues its oversight of PHMSA's enforcement activities, and as it considers new proposals to increase compliance with federal pipeline safety regulations, Congress may evaluate how PHMSA's authorities to set standards, assess penalties, and directly affect pipeline operations may reinforce one another to improve U.S. pipeline safety.

Regulation of Canadian Oil/Tar Sands Crude Pipelines

Canadian oil exports to the United States have been increasing rapidly, primarily due to growing output from the oil sands in Western Canada.[151] Oil sands (also referred to as tar sands) are a mixture of clay, sand, water, and heavy black viscous oil known as bitumen. Oil sands are processed to extract the bitumen, which can then be upgraded into a product that is suitable for pipeline transport. Canada's oil sands production can be exported as either a light, upgraded synthetic crude ("syncrude") or a heavy crude oil that is a blend of bitumen diluted with lighter hydrocarbons ("dilbit") to ease transport. The bulk of oil sands' supply growth is expected to be in the form of the latter.[152] Five major pipelines have been constructed in recent years to link the oil sands region to markets in the United States. A sixth pipeline, Keystone XL, was rejected in January 2012 by the U.S. State Department, although the developer plans to reapply for a federal permit with a modified route.[153] If ultimately approved and constructed, Keystone XL would bring Canada's total U.S. petroleum export capacity to over 4.1 million barrels per day, enough capacity to carry over a third of current U.S. petroleum imports.[154]

This expansion of petroleum pipelines from Canada has generated considerable controversy in the United States. One specific area of concern has been perceived new risks to pipeline integrity of transporting heavy Canadian crudes. Some opponents of the new Canadian oil pipelines, notably the Natural Resources Defense Council (NRDC), have argued that these pipelines could be more likely to fail and cause environmental damage than other crude oil pipelines because the bitumen mixtures they would carry are "significantly more corrosive to pipeline systems than conventional crude," among other reasons.[155] NRDC has called for a moratorium on approving new oil pipelines from oil sands regions, and a review of existing pipeline permits, until these safety concerns are researched further and addressed in federal environmental and safety studies. Canadian officials and other stakeholders have rejected these arguments, however, citing factual inaccuracies and a flawed methodology in the NRDC analysis, which compares pipeline spill rates in Canada to those in the United States.[156]

Some in Congress have called for a review of PHMSA regulations to determine whether new regulations for Canadian heavy crudes are needed to account for any unique properties they may have. Accordingly, P.L. 112-90 requires PHMSA to review whether current regulations are sufficient to regulate pipelines transmitting "diluted bitumen," and analyze whether such oil presents an increased risk of release (§16). This study, which is being performed by the Transportation Research Board of the National Academy of Sciences, has not yet been completed.

Pipeline Security Regulations

As noted earlier in this report, federal pipeline security activities to date have relied upon voluntary industry compliance with PHMSA security guidance and TSA security best practices. By initiating this voluntary approach, PHMSA sought to speed adoption of security measures by industry and avoid the publication of sensitive security information (e.g., critical asset lists) that would normally be required in public rulemaking.[157] Provisions in P.L. 109-468 require the DOT Inspector General to "address the adequacy of security standards for gas and oil pipelines" (§23(b)(4)). P.L. 110-53 similarly directs TSA to promulgate pipeline security regulations and carry out necessary inspection and enforcement—if the agency determines that regulations are appropriate (§1557(d)). Addressing this issue, the 2008 IG report states that

> TSA's current security guidance is not mandatory and remains unenforceable unless a regulation is issued to require industry compliance.... PHMSA and TSA will need to conduct covert tests of pipeline systems' vulnerabilities to assess the current guidance as well as the operators' compliance.[158]

Although TSA's FY2005 budget justification stated that the agency would "issue regulations where appropriate to improve the security of the [non-aviation transportation] modes," the agency has not done so for pipelines, and it is not currently working on such regulations. The pipeline industry has long expressed concern that new security regulations and related requirements may be "redundant" and "may not be necessary to increase pipeline security."[159] The PHMSA Administrator, in 2007, similarly testified that enhancing security "does not necessarily mean that we must impose regulatory requirements."[160] TSA officials have questioned the IG assertions regarding pipeline security regulations, particularly the IG's call for covert testing of pipeline operator security measures. They have argued that the agency is complying with the letter of P.L. 110-53 and that its pipeline operator security reviews are more than paper reviews.[161] In accordance with P.L. 110-53 (§1557 (b)), TSA has been implementing a multi-year program of pipeline system inspections, including documentation of findings and follow up reviews.[162] Because the TSA believes the most critical U.S. pipeline systems generally meet or exceed industry security guidance, the agency believes it achieves better security with voluntary guidelines, and maintains a more cooperative and collaborative relationship with its industry partners as well.[163]

Although the TSA believes its voluntary approach to pipeline security is adequate, Canadian pipeline regulators have come to a different conclusion. In 2010 the National Energy Board of Canada mandated security regulations for jurisdictional Canadian petroleum and natural gas pipelines, some of which are cross-border pipelines serving export markets in the United States. A number of companies operate pipelines in both countries. In announcing these new regulations, the board stated that it had considered adopting the existing security standards "as guidance" rather than an enforceable standard, but "taking into consideration the critical importance of energy infrastructure protection," the board decided to adopt the standard into the regulations.[164] Establishing pipeline security regulations in Canada is not completely analogous to doing so in the United States, as the Canadian pipeline system is much smaller and operated by far fewer companies than the U.S. system. Nonetheless, Canada's choice to regulate pipeline security may raise questions as to why the United States has not. In its oversight of potential pipeline security regulations, Congress may evaluate the effectiveness of the current voluntary pipeline security standards based on findings from the TSA's CSR reviews, pipeline inspections, DHS cybersecurity alerts, and any future DOT Inspector General reports.

Additional Issues

In addition to the issues mentioned above, Congress may consider several issues related to proposed legislation or otherwise raised by pipeline stakeholders.

Accuracy and Completeness of Pipeline System Records

On January 3, 2011, as a response to its initial investigation of the San Bruno pipeline accident, the NTSB issued urgent new safety recommendations "to address record-keeping problems that could create conditions in which a pipeline is operated at a higher pressure than the pipe was built to withstand."[165] The NTSB issued these recommendations after it had concluded that there were significant errors in the records characterizing the San Bruno pipeline, and that "other pipeline operators may have discrepancies in their records that could potentially compromise the safe operation of pipelines throughout the United States."[166] PHMSA officials have also testified that some operators may not be collecting all the pipeline system data necessary to fully evaluate safety and compliance with federal regulations.[167] In 2006, questions were raised about the accuracy of pipeline location data provided by operators and maintained by PHMSA in the National Pipeline Mapping System (NPMS).[168] At the time, agency officials reportedly acknowledged limitations in NPMS accuracy, but did not publicly discuss plans to address them. P.L. 112-90 authorizes PHMSA to collect additional geospatial and technical data from pipeline operators to achieve the purposes of the NPMS (§11(a) and §12). Congress may review whether these or other statutory measures are sufficient to verify that pipeline operator information is complete and correct, particularly for older parts of the pipeline network.

Mandatory Internal Inspection or Hydrostatic Testing

Some proposals would increase requirements for pipeline operators to conduct internal inspections of transmission pipelines using "smart pigs," robotic devices sent through pipelines to take physical measurements continuously along the way.[169] In its San Bruno accident investigation report, the NTSB has recommended that all natural gas transmission pipelines be configured to accommodate such internal inspection tools. However, experts note that there are different pipeline inspection techniques with overlapping capabilities and different strengths.[170] While an effective technology for detecting corrosion in many applications, smart pigs have limitations as a general tool for assessing the integrity of pipelines. For example, although smart pigs may be good corrosion detectors, they are still a developing technology and may be somewhat less effective in detecting other types of pipeline anomalies (e.g., cracks). Operators also maintain that smart pigging may be less useful for predicting future problems with pipeline integrity than other federally approved maintenance techniques like "direct assessment" (49 C.F.R. 192.903) wherein pipelines are examined externally based on risk data and other factors.[171] Furthermore, because many older pipelines contain sharp turns and other obstructions due to historical construction techniques, they cannot accommodate smart pig devices without significant and costly pipeline modifications to make them more "piggable." Consequently, some industry stakeholders caution against unrealistic expectations for the capabilities of smart pigs as a stand-alone pipeline inspection tool.[172]

As an alternative to internal inspection where such inspection cannot currently be performed, some policy makers have called for mandatory hydrostatic testing of pipelines to verify their integrity. Hydrostatic testing involves filling a pipeline with water under pressure greater than the anticipated operating pressure to determine if it is structurally sound and does not leak. Such testing is common for new pipelines that have not yet entered service. Because it uses only water, hydrostatic testing poses relatively little direct risk to the public or the environment, but when used for operating pipelines it necessarily interrupts pipeline service. Injecting water into pipelines is also costly, and may create safety problems since water is corrosive and may be difficult to remove completely from a pipeline once testing is completed.[173] Nonetheless, as noted above, the NTSB has recommended that all natural gas transmission pipelines constructed before 1970 be subjected to hydrostatic pressure tests. P.L. 112-90 requires verification of maximum allowable operating pressure for all natural gas transmission pipelines "as expeditiously as economically feasible" (§23(a)). The act also authorizes regulations for pressure verification that "shall consider ... pressure testing; and ... other alternative methods, including in-line inspections" (§23(a)). As Congress examines any new federal requirements for pipeline inspection, it may consider smart pig devices and hydrostatic testing as only two techniques in a portfolio of maintenance practices operators may need to employ to ensure their pipelines are physically sound.

Emergency Response Plan Disclosure

Federal regulations require pipeline operators to prepare emergency response plans for pipeline spills and to make those plans available for inspection by PHMSA and local emergency response agencies (49 C.F.R. 192.605). Some stakeholders have proposed that these plans also be made available to the public to allow for additional review of their adequacy and to provide better risk and response information to people living near pipelines.[174] Operators reportedly have resisted such disclosures on the grounds that their emergency response plans contain confidential customer and employee information.[175] They also raise concerns that the plans contain security-sensitive information about pipeline vulnerabilities and spill scenarios which could be useful to terrorists.[176] P.L. 112-90 requires PHMSA to collect and maintain copies of pipeline emergency plans for public availability excluding any proprietary or security-sensitive information (§6(a)). As oversight of this issue continues, Congress may consider the tradeoffs between public awareness and pipeline security in a general operating environment where both safety and security hazards may be significant.

Pipeline Water Crossings

The 2011 oil spill into the Yellowstone River near Laurel, MT, appears to have been the result of the buried oil pipeline becoming exposed due to scouring of the river bottom during unusually heavy flooding.[177] Prior to the flooding, a depth-of-cover survey by the operator verified that the pipeline was at least five feet below the riverbed, exceeding a four-foot minimum cover requirement in PHMSA regulations.[178] Because the four-foot requirement appears to have been insufficient to prevent riverbed pipeline exposure in this case, some policy makers have called for a review of pipeline river crossings and associated safety requirements nationwide. P.L. 112-90 mandates a review of the adequacy of PHMSA regulations with respect to pipelines that cross inland bodies of water at least 100 feet wide and, based on the review's findings, requires PHMSA to develop legislative recommendations for changing existing regulations (§28(a)). The agency has not yet released this study.

CONCLUSION

Both government and industry have taken numerous steps to improve pipeline safety and security over the last 10 years. While stakeholders across the board agree that federal pipeline safety programs have been on the right track, major pipeline incidents since 2010 suggest there continues to be significant room for improvement. Likewise, the threat of physical and cyberattack on U.S. pipeline infrastructure remains a concern. The NTSB has identified improvement of federal pipeline safety oversight as a "top ten" priority for 2013. The leading pipeline industry associations have concurred. The American Gas Association has expressed support for these NTSB recommendations, stating that "pipeline safety and integrity is the top priority for the natural gas industry."[179] The Interstate Natural Gas Association of America (INGAA) has stated that "INGAA members are addressing all of the issues that the NTSB has outlined ... and more."[180] The Association of Oil Pipe Lines has also welcomed the NTSB's focus on pipeline safety, stating that "operators are hard at work on safety improvements suggested by NTSB, PHMSA and their own initiatives."[181] Whether the renewed efforts by industry, combined with additional oversight by federal agencies, will further enhance the safety and security of U.S. pipelines remains to be seen.

As Congress oversees the federal pipeline safety program and the federal role in pipeline security, key issues of focus may be pipeline agency staff resources, automatic pipeline shutoff valves, penalties for safety violations, safety regulations for oil sands crudes, and the possible need for pipeline security regulations, among other concerns. In addition to these specific issues, Congress may assess how the various elements of U.S. pipeline safety and security activity fit together in the nation's overall strategy to protect transportation infrastructure. Pipeline safety and security necessarily involve many groups: federal agencies, oil and gas pipeline associations, large and small pipeline operators, and local communities. Reviewing how these groups work together to achieve common goals could be an oversight challenge for Congress.

End Notes

[1] Hazardous liquids primarily include crude oil, gasoline, jet fuel, diesel fuel, home heating oil, propane, and butane. Other hazardous liquids transported by pipeline include anhydrous ammonia, carbon dioxide, kerosene, liquefied ethylene, and some petrochemical feedstocks.

[2] Pipeline and Hazardous Materials Safety Administration, "Annual Report Mileage for Hazardous Liquid or Carbon Dioxide Systems," online table, October 31, 2012, http://www.phmsa.dot.gov/pipeline/library/data-stats.

[3] Richard A. Rabinow, "The Liquid Pipeline Industry in the United States: Where It's Been, Where It's Going," Prepared for the Association of Oil Pipe Lines, April 2004, p. 4.

[4] Pipeline and Hazardous Materials Safety Administration, "Annual Report Mileage for Natural Gas Transmission & Gathering Systems," online table, October 31, 2012, http://www.phmsa.dot.gov/pipeline/library/data-stats. There are also approximately 6,300 miles of offshore gathering pipelines. Gathering pipelines in the Outer Continental Shelf regulated by the Department of the Interior are outside the scope of this report.

[5] Ibid.

[6] Energy Information Administration, "About U.S. Natural Gas Pipelines," June 2007, pp. 1, 29, http://www.eia.doe.gov/pub/oil_gas/natural_gas/analysis_publications/ngpipeline/fullversion.pdf.

[7] Pipeline and Hazardous Materials Safety Administration, "Annual Report Mileage for Gas Distribution Systems," online table, October 31, 2012, http://www.phmsa.dot.gov/pipeline/library/data-stats.

[8] Pipeline and Hazardous Materials Safety Administration, "Annual Report Mileage for Liquefied Natural Gas (LNG) Facilities," November 29, 2012, http://www.phmsa.dot.gov/pipeline/library/data-stats.

[9] Bureau of Transportation Statistics, "Table 2-1: Transportation Fatalities by Mode," online table, April 2012, http://www.bts.gov/publications/national_transportation_statistics/html/table_02_01.html.

[10] Pipeline and Hazardous Materials Safety Administration, "Significant Pipeline Incidents," web page, December 12, 2012, http://primis.phmsa.dot.gov/comm/reports/safety/ SigPSI.html?nocache=6256.

[11] Ibid.

[12] J.L. Shreeve, "Science & Technology: The Enemy Within," The Independent. London, UK, May 31, 2006, p. 8.

[13] Government Accountability Office (GAO), Security Assistance: Efforts to Secure Colombia's Caño Limón-Coveñas Oil Pipeline Have Reduced Attacks, but Challenges Remain, GAO-05-971, September 2005, p. 15; Stratfor Forecasting, Inc.," Colombia: The FARC's Low-Level Pipeline Campaign," Stratfor Today, June 23, 2008. http://www.stratfor.com/analysis/colombia_farcs_low_level_pipeline_campaign?ip_auth_redirect=1.

[14] President's Commission on Critical Infrastructure Protection, Critical Foundations: Protecting America's Infrastructures, Washington, DC, October 1997.

[15] Katherine Houreld, "Militants Say 3 Nigeria Pipelines Bombed," Associated Press, May 8, 2007.

[16] Reed Johnson, "Six Pipelines Blown Up in Mexico," Los Angeles Times, September 11, 2007. p. A-3.

[17] U.S. Department of Justice, "Four Individuals Charged in Plot to bomb John F. Kennedy International Airport," Press release, June 2, 2007.

[18] Ben Gelinas, "New Letter Threatens Resumption of 'Action' against B.C. Pipelines," Calgary Herald, April 15, 2010.

[19] Steve Fainaru and William Booth, "Mexico's Drug Cartels Siphon Liquid Gold," Washington Post, December 13, 2009.

[20] "Already Hard at Work on Security, Pipelines Told of Terrorist Threat," Inside FERC, McGraw-Hill Companies, January 3, 2002.

[21] Alyeska Pipeline Service Co., "Overview of TAPS." Internet page, June 27, 2012, http://www.alyeska-pipe.com/ TAPS.

[22] David S. Cloud, "A Former Green Beret's Plot to Make Millions Through Terrorism," Ottawa Citizen, December 24, 1999, p. E15.

[23] Y. Rosen, "Alaska Critics Take Potshots at Line Security," Houston Chronicle, February 17, 2002.

[24] Wesley Loy, "Web Post Urges Jihadists to Attack Alaska Pipeline," Anchorage Daily News, January 19, 2006.

[25] U.S. Attorney's Office, Middle District of Pennsylvania, "Man Convicted of Attempting to Provide Material Support to Al-Qaeda Sentenced to 30 Years' Imprisonment," Press release, November 6, 2007; A. Lubrano and J. Shiffman, "Pa. Man Accused of Terrorist Plot," Philadelphia Inquirer, February 12, 2006, p. A1.

[26] Transportation Security Administration, Office of Intelligence, Pipeline Threat Assessment, January 18, 2011, p. 3.

[27] Carol Cratty, "Man Accused in Attempted Bombing of Oklahoma Gas Pipeline," CNN, August 12, 2011.

[28] "Grand Jury Indicts Plano Gas Pipeline Bomb Suspect on Weapons Charge," Associated Press, July 11, 2012.

[29] For a more comprehensive analysis of U.S. pipeline cybersecurity issues, see CRS Report R42660, Pipeline Cybersecurity: Federal Policy, by Paul W. Parfomak.

[30] "Spear-phishing" involves sending official-looking e-mails to specific individuals to insert harmful software programs (malware) into protected computer systems; to gain unauthorized access to proprietary business information; or to access confidential data such as passwords, social security numbers, and private account numbers.

[31] Industrial Control Systems Cyber Emergency Response Team (ICS-CERT), "Gas Pipeline Cyber Intrusion Campaign," ICS-CERT Monthly Monitor, April 2012, p.1, http://www.us-cert.gov/control_systems/pdf/ICSCERT_Monthly_Monitor_Apr2012.pdf.

[32] McAfee Foundstone Professional Services and McAfee Labs, Global Energy Cyberattacks: "Night Dragon," white paper, February 10, 2011, p. 3, http://www.mcafee.com/us/resources /white-papers/wp-global-energy-cyberattacksnight-dragon.pdf.

[33] Tobias Walk, 2007, p. 7.

[34] Eric Byres, February 2012.

[35] Safety and security of liquefied natural gas (LNG) facilities used in gas pipeline transportation is regulated under CFR Title 49, Part 193.

[36] Pipeline and Hazardous Material Safety Administration (PHMSA), personal communication, December 13, 2012.

[37] 49 U.S.C. 60107.

[38] PHMSA, December 13, 2012.

[39] 49 U.S.C. 60125.

[40] U.S. Office of Management and Budget, Budget of the United States Government, Fiscal Year 2013: Appendix February 13, 2012.

[41] Pipeline and Hazardous Material Safety Administration (PHMSA), "PHMSA Pipeline Safety Program: Summary of Enforcement Actions," web page, December 14, 2012, http://primis. phmsa.dot.gov/comm/reports/enforce/ Actions_opid_0.html?nocache=8828.

[42] Pipeline and Hazardous Material Safety Administration (PHMSA), "PHMSA Pipeline Safety Program: Summary of Cases Involving Civil Penalties," web page. December 14, 2012, http://primis.phmsa.dot.gov/comm/reports/enforce/ CivilPenalty_opid_0.html?nocache=9288#_TP_1_tab_1.

[43] Presidential Decision Directive 63, Protecting the Nation's Critical Infrastructures, May 22, 1998.

[44] In November 2004, the President signed the Norman Y. Mineta Research and Special Programs Improvement Act (P.L. 108-426), which eliminated RSPA and placed the Office of Pipeline Safety under the new Pipeline and Hazardous Material Safety Administration. This restructuring did not significantly alter the authorities or activities of the OPS.

[45] Research and Special Programs Administration (RSPA), RSPA Pipeline Security Preparedness, December 2001.

[46] Ellen Engleman, Administrator, Research and Special Programs Administration (RSPA), statement before the Subcommittee on Energy and Air Quality, House Energy and Commerce Committee, March 19, 2002.

[47] Ellen Engleman, Administrator, Research and Special Programs Administration (RSPA), statement before the Subcommittee on Highways and Transit, House Transportation and Infrastructure Committee, February 13, 2002.

[48] James K. O'Steen, Research and Special Programs Administration (RSPA), Implementation of RSPA Security Guidance, presentation to the National Association of Regulatory Utility Commissioners, February 25, 2003.

[49] Office of Pipeline Safety (OPS), personal communication, June 10, 2003.

[50] P.L. 107-355 encourages the implementation of state "one call" excavation notification programs (§2) and allows states to enforce "one-call" program requirements. The act expands criminal responsibility for pipeline damage to cases where damage was not caused "knowingly and willfully" (§3). The act adds provisions for ending federal-state pipeline oversight partnerships if states do not comply with federal requirements (§4).

[51] A 2006 Government Accountability Office (GAO) report found that PHMSA's gas integrity management program benefitted public safety, although the report recommended revisions to PHMSA's performance measures. See GAO, "Natural Gas Pipeline Safety: Integrity

Management Benefits Public Safety, but Consistency of Performance Measures Should Be Improved," GAO-06-946, September 8, 2006, pp. 2-3.

[52] PHMSA issued final regulations requiring operators of natural gas distribution pipelines to adopt integrity management programs similar to existing requirements for gas transmission pipelines on December 4, 2009.

[53] HSPD-7 supersedes PDD-63 (par. 37).

[54] Executive Order 13416, "Strengthening Surface Transportation Security," December 5, 2006.

[55] The PSD was formerly known as the Pipeline Security Program Office. The OSPIE was previously known as the Office of Transportation Sector Network Management and, before that, the Intermodal Security Program Office.

[56] Transportation Security Administration, Pipeline Modal Annex, June 2007, pp. 10-11. G8 = Group of Eight (the United States, the United Kingdom, Canada, France, Germany, Italy, Japan, and Russia).

[57] Department of Transportation (DOT), "Action Taken and Actions Needed to Improve Pipeline Safety," CC-2004- 061, June 16, 2004, p. 21.

[58] Government Accountability Office (GAO), Pipeline Security: TSA Has Taken Actions to Help Strengthen Security, but Could Improve Priority-Setting and Assessment Processes, GAO-10-867, August, 2010, Executive Summary.

[59] Jack Fox, General Manager, Pipeline Security Division, Transportation Security Administration (TSA), personal communication, July 25, 2012.

[60] Mike Gillenwater, TSA, "Pipeline Security Overview," presented to the Alabama Public Service Commission Gas Pipeline Safety Seminar, Montgomery, AL, December 11, 2007.

[61] Department of Homeland Security, "Extension of Agency Information Collection Activity Under OMB Review: Critical Facility Information of the Top 100 Most Critical Pipelines," 76 Federal Register 62818, October 11, 2011.

[62] Transportation Security Administration, personal communication, February 24, 2012.

[63] TSA, TSA Multi-Modal Criticality Evaluation Tool, TSA Threat Assessment and Risk Management Program, slide presentation, April 15, 2003.

[64] TSA, Intermodal Security Program Office, Pipeline Security Best Practices, October 19, 2005, p. 1.

[65] Transportation Security Administration, Pipeline Security Guidelines, April, 2011, pp. 16-19; Personal communication, February 2, 2010.

[66] U.S. Office of Management and Budget, Budget of the United States Government, Fiscal Year 2012: Appendix, February 2012, p. 508.

[67] Jack Fox, personal communication, July 25, 2012.

[68] Kip Hawley, Assistant Secretary, Department of Homeland Security, Testimony before the Senate Committee on Commerce, Science, and Transportation hearing on Federal Efforts for Rail and Surface Transportation Security, January 18, 2007.

[69] Pipeline and Hazardous Materials Safety Administration (PHMSA), "Pipeline Safety: Lessons Learned From a Security Breach at a Liquefied Natural Gas Facility," Docket No. PHMSA-04-19856, Federal Register, Vol. 71, No. 249, December 28, 2006, p. 78269; TSA, Intermodal Security Program Office, personal communication, August 30, 2006.

[70] Transportation Security Administration, personal communication, July 6, 2007.

[71] Government Accountability Office, GAO Watchdog, "Transportation Security's Efforts To Ensure Pipeline Security," Assignment No. 440768, Internet database, February 4, 2010.

[72] U.S. Government Accountability Office (GAO), Pipeline Security: TSA Has Taken Actions to Help Strengthen Security, but Could Improve Priority-Setting and Assessment Processes, GAO-10-867, August, 2010, pp. 54-55.

[73] Jerald E. Levine, Director, Departmental GAO/OIG Liaison Office, Department of Homeland Security, Letter to GAO, July 23, 2010.
[74] U.S. Code of Federal Regulations. 18 C.F.R. 157.
[75] Federal Energy Regulatory Commission (FERC), personal communication, May 22, 2003.
[76] Federal Energy Regulatory Commission (FERC), News release, R-01-38, Washington, DC, September 14, 2001.
[77] Federal Energy Regulatory Commission (FERC), Federal Energy Regulatory Commission Annual Report FY2005, 2006, p. 19. These are the most recent specific figures reported.
[78] Federal Energy Regulatory Commission (FERC), Federal Energy Regulatory Commission Annual Report FY2006, 2007, p. 23.
[79] Federal Energy Regulatory Commission (FERC), news release, R-03-08, Washington, DC. February 20, 2003.
[80] Christian Schmollinger, "FERC OKs Emergency Reconstruction," Natural Gas Week, May 13, 2003.
[81] National Transportation Safety Board, 2011 Annual Report, NTSB/SPC-12/04, p. iii.
[82] National Transportation Safety Board, "NTSB Most Wanted List: Enhance Pipeline Safety," November 2012.
[83] National Transportation Safety Board (NTSB), "Public Hearing: Natural Gas Pipeline Explosion and Fire, San Bruno, CA, September 9, 2010," web page, March 15, 2011, http://www.ntsb.gov/Events/2011/San_Bruno_CA/ default.htm.
[84] National Transportation Safety Board (NTSB), "Pacific Gas and Electric Company Natural Gas Transmission Pipeline Rupture and Fire, San Bruno, CA, September 9, 2010," NTSB/PAR-11/01, August 30, 2011, p.118.
[85] Deborah A.P. Hersman, Chairman, National Transportation Safety Board, "Opening Statement, Pipeline Accident Report – San Bruno, California, September 9, 2010," August 30, 2011.
[86] NTSB, August 30, 2011, p. 123.
[87] Generally, Class 3 locations have 46 or more buildings intended for human occupancy or lie within 100 yards of either a building or outside area of public assembly; Class 4 locations are areas where buildings with four or more stories are prevalent. For precise definitions, see 49 C.F.R. 192.5.
[88] NTSB, August 30, 2011, pp. 128-132.
[89] U.S. Office of Management and Budget, February 2012, p. 1046.
[90] John D. Porcari, Deputy Secretary, U.S. Department of Transportation, Testimony before the House Committee on Transportation and Infrastructure, Hearing on the Enbridge Pipeline Oil Spill in Marshall, MI, September 15, 2010.
[91] Linda Daugherty, Pipeline and Hazardous Materials Safety Administration, Testimony before the National Transportation Safety Board hearing "Natural Gas Pipeline Explosion and Fire, San Bruno, CA, September 9, 2010," March 2, 2011.
[92] Linda Daugherty, Pipeline and Hazardous Materials Safety Administration, personal communication, December 13, 2012.
[93] Linda Daugherty, December 13, 2012.
[94] Pipeline and Hazardous Materials Safety Administration, personal communication, November 4, 2010.
[95] U.S. Department of Transportation, Pipeline And Hazardous Materials Safety Administration, FY 2011 Budget Request, February 1, 2010, p. 31, http://www.dot.gov/budget/ 2011/budget estimates/phmsa.pdf.
[96] Linda Daugherty, December 13, 2012.

[97] National Association of Pipeline Safety Representatives, RE: Request for Waiver of Prior Three Year Average State Expense Component of the Pipeline Safety Grant Program, letter to Mr. Jeffrey D. Weise, Associate Administrator for Pipeline Safety, Pipeline And Hazardous Materials Safety Administration, October 15, 2010, http://www.wutc.wa.gov/webimage.nsf/web+objects/CCOPS_DOCs_by_Year/$file/NAPSR%20letter%20to%20PHM SA%20dated%2010-15-09-Waiver%20Request.pdf.

[98] Stephanie Seay, "Budget Woes May Impede Local Pipe Safety Efforts," Gas Daily, November 8, 2010.

[99] The Honorable Cynthia Quarterman, Administrator, Pipeline and Hazardous Materials Safety Administration, Remarks at the "Different Pathways to a Common Goal: PIPA, Damage Prevention, & Greater Public Awareness and Involvement Conference," Pipeline Safety Trust, New Orleans, LA, November 5, 2010.

[100] Senator Daniel K. Inouye, opening statement before the Senate Committee on Commerce, Science and Transportation, hearing on the President's FY2006 Budget Request for the Transportation Security Administration (TSA), February 15, 2005.

[101] Congressman Gus M. Bilirakis, Remarks before the House Committee on Homeland Security, Subcommittee on Management, Investigations, and Oversight hearing on "Unclogging Pipeline Security: Are the Lines of Responsibility Clear?," Plant City, FL, April 19, 2010.

[102] Department of Homeland Security, "Intent to Request Approval from OMB of One New Public Collection of Information: Pipeline Corporate Security Review," 74 Federal Register 42086, August 20, 2009.

[103] U.S. Government Accountability Office, Transportation Security: Comprehensive Risk Assessments and Stronger Internal Controls Needed to Help Inform TSA Resource Allocation, GAO-09-492, March 2009, p. 30, http://www.gao.gov/new.items/d09492.pdf.

[104] For example, see Hon. William J. Pascrell, Jr., statement at the House Committee on Transportation and Infrastructure, Subcommittee on Highways, Transit and Pipelines, hearing on Pipeline Safety, March 16, 2006.

[105] TSA, Pipeline Security Division, personal communication, November 5, 2010.

[106] Congressman Gus M. Bilirakis, April 19, 2010.

[107] The committee print is available at http://homeland.house.gov/sites/homeland.house.gov/files/documents/pdf/ 091411_TS_Subcommittee%20Print.pdf.

[108] John Upton, "Gas Fueled Blaze for Almost Two Hours," San Francisco Examiner, September 15, 2010.

[109] National Transportation Safety Board, Texas Eastern Transmission Corporation Natural Gas Pipeline Explosion and Fire, Edison, New Jersey, March 23, 1994, NTSB/PAR-95/01, January 18, 1995.

[110] U.S. Department of Transportation, Remotely Controlled Valves on Interstate Natural Gas Pipelines, September 1999, p. 22.

[111] U.S. Department of Transportation, September 1999, pp. 23-24.

[112] Rich Connell, John Hoeffel and Marc Lifsher, "Lawmakers Move to Impose New Requirements for Pipeline Shutoff Valves," Los Angeles Times, September 14, 2010.

[113] Christina Sames, Vice President, American Gas Association, Remarks at the Different Pathways to a Common Goal: PIPA, Damage Prevention, & Greater Public Awareness and Involvement Conference, Pipeline Safety Trust, New Orleans, LA, November 4, 2010.

[114] National Transportation Safety Board (NTSB), Pipeline Special Investigation Report: Evaluation of Accident Data and Federal Oversight of Petroleum Product Pipelines, NTSB/SIR-96/02, 1996, p. 37.

[115] Ibid.

[116] Ibid. p. 39.

[117] 49 CFR 195.452(i)(4)

[118] Pacific Gas and Electric Company, "Re: Updates on Natural Gas Transmission System," Letter to the California Public Utilities Commission, October 25, 2010, p. 2-2, http://www.pge.com/includes/docs/pdfs/about/newsroom/ puc_updates_oct252010.pdf.

[119] Cecil R. Sparks, Thomas R. Morrow, and John P. Harrell, "Cost Benefit Study of Remote Controlled Main Line Valves, Final Report to Gas Research Institute," Southwest Research Institute, San Antonio, TX, Report No. GRI- GRI-98/0076, May 1998.

[120] U.S. Department of Transportation, September 1999, p. 11.

[121] Pacific Gas and Electric Company, October 25, 2010, p. 2-3.

[122] John Upton, "PG&E Faces High Costs on Pipelines," New York Times, March 3, 2011.

[123] Oak Ridge National Laboratory, Studies for the Requirements of Automatic and Remotely Controlled Shutoff Valves on Hazardous Liquids and Natural Gas Pipelines with Respect to Public and Environmental Safety, ORNL/TM2012/411, October 31, 2012, p. xxviii.

[124] Ibid.

[125] Charles H. Batten, "Engineering Analysis of Olympic Pipe Line Company's Safety and Risk Mitigation Features for Application No. 96-1 Cross Cascade Pipeline Project," Batten & Associates, Inc., Locust Grove, VA, February 8, 1999, p. 19, http://www.efsec.wa.gov/oplarchive/oplpft/Cca/chb-3.pdf.

[126] National Transportation Safety Board, Pipeline Accident Summary Report: Natural Gas Pipeline Rupture and Fire During Dredging of Tiger Pass, Louisiana, October 23, 1996, PR98-916501, NTSB/PAR-98/01/SUM, September 28, 1998, p. 15.

[127] See, for example: Tyler Williams, "Cyber Security Threats to Pipelines and Refineries," Pipeline & Gas Journal, November 1, 2007.

[128] Pipeline and Hazardous Materials Safety Administration, Leak Detection Study—DTPH56-11-D-000001, Final Report No. 12-173, prepared by Kiefner & Associates, Inc., Worthington, OH, December 10, 2012.

[129] U.S. Department of Transportation, September 1999, pp. 19-20.

[130] William M. Nugent, First Vice President, National Association of Regulatory Utility Commissioners, testimony before the Senate Energy and Natural Resources Committee hearing on Federal, State, and Local Impediments to Siting Energy Infrastructure, May 15, 2001.

[131] National Commission on Energy Policy, Siting Critical Energy Infrastructure: An Overview of Needs and Challenges, Washington, DC, June 2006, p. 1.

[132] Energy Information Administration, Additions to Capacity on the U.S. Natural Gas Pipeline Network: 2005, August 2006. p. 11.

[133] Federal Energy Regulatory Commission (FERC), "Commission Approves Revised $1 Billion Millennium Pipeline Project to Bring New Gas Service to the Northeast," press release, December 21, 2006. See, for example: Randal C. Archibold, "Fighting Plans for a Gas Pipeline: Not Under My Backyard," New York Times, August 7, 2001.

[134] See, for example, Samantha Santa Maria, "Energy Projects: Rockies Express Add-on Pipe Projects Face Several Obstacles to Building in US Northeast," Inside F.E.R.C., October 22, 2007.

[135] U.S. Department of Transportation, "DOT Issues Much-Anticipated Rules to Enhance Pipeline Safety," Office of Public Affairs, press release, December 3, 2009.

[136] See, for example: Representative James L. Oberstar, Statement before the House Committee on Transportation and Infrastructure, Hearing on the Enbridge Pipeline Oil Spill in Marshall, MI, September 15, 2010.

[137] S.L. Gerard, Pipeline and Hazardous Materials Administration, testimony before the House Energy and Commerce Committee, Energy and Air Quality Subcommittee hearing on Pipeline Safety, Serial No. 109-84, April 27, 2006, p. 14.

[138] Ibid. These figures only reflect administrative enforcement cases. They exclude cases that PHMSA has referred to the Department of Justice for civil and criminal enforcement under 49 CFR 190.231 and 190.235.

[139] Pipelines and Hazardous Materials Safety Administration, "Civil Penalty Cases: Nationwide," web page, February 8, 2012, http://primis.phmsa.dot.gov/comm/reports/ enforce/Civil Penalty_opid_0.html?nocache=9398; "Colorado Pipeline Company Fined 2.3 Million After Explosion," Clean Skies News, December 1, 2009.

[140] PHMSA, February 8, 2012.

[141] Pipeline and Hazardous Materials Safety Administration, Consent Agreement: In the Matter of Kinder Morgan Energy Partners, L.P., Respondent, CPF No. 5-2005-5025H, April 4, 2006.

[142] Kinder Morgan Energy Partners, L.P., "Kinder Morgan Energy Partners Enters into Consent Agreement with PHMSA," press release, Houston, TX, April 10, 2006.

[143] U.S. Environmental Protection Agency, "Kinder Morgan, SFPP Agree to Pay Nearly $5.3 Million to Resolve Federal And State Environmental Violations," press release, May 21, 2007.

[144] U.S. Environmental Protection Agency, "Plains Pipeline to Spend $41 Million to Prevent Oil Spills Across 10,000 Miles of Pipeline," press release, August 10, 2010.

[145] Enbridge Energy Partners, L.P., Enbridge Energy Partners, L.P. Third Quarter 2010 Earnings, Slide presentation, October 28, 2010, p. 8, http://phx.corporate-ir.net/External. File?item= UGFyZW50SUQ9MjY2NzE3N3xDaGlsZElEPTQwMTI5MXxUeXBlPTI=&t=1.

[146] Enbridge Energy Partners, L.P., October 28, 2010, p. 8.

[147]"Olympic Pipe Line, Others Pay Out Record $75 Million in Pipeline Explosion Wrongful Death Settlement," Business Wire, April 10, 2002.

[148] National Transportation Safety Board, Pipeline Accident Report, PAR-03-01, February 11, 2003.

[149] El Paso Corp., Quarterly Report Pursuant to Section 13 or 15(d) of the Securities Exchange Act of 1934, Form 10- Q, for the period ending June 30, 2002, Houston, TX, 2002.

[150] El Paso Corp., 2002.

[151] For further discussion and analysis of Canadian oil sands pipeline issues, see CRS Report R41875, The U.S.-Canada Energy Relationship: Joined at the Well, by Paul W. Parfomak and Michael Ratner.

[152] Canadian Association of Petroleum Producers (CAPP), "Crude Oil: Forecast, Markets & Pipelines," Calgary, AB, June 2010, p. 7, http://www.capp.ca/GetDoc.aspx?DocId=173003; CRS analysis.

153 See CRS Report R41668, Keystone XL Pipeline Project: Key Issues, by Paul W. Parfomak et al.; U.S. Department of State, "Briefing on the Keystone XL Pipeline," briefing transcript, January 18, 2012, http://www.state.gov/r/pa/prs/ ps/2012/01/181492.htm; TransCanada Corp., "TransCanada Will Re-Apply for a Keystone XL Permit," press release, January 18, 2012.

[154] TransCanada has proposed a pipeline spur from the Keystone XL pipeline to the Bakken oil shale field in Montana, North Dakota, and South Dakota, known as the Bakken Marketlink Project.

[155] Anthony Swift, Susan Casey-Lefkowitz, and Elizabeth Shope, "Tar Sands Pipelines Safety Risks," Natural Resources Defense Council, February 2011, p. 6.

[156] Canadian Energy Resources Conservation Board, "ERCB Addresses Statements in Natural Resources Defense Council Pipeline Safety Report," press release, Calgary, Alberta, February 16, 2011.

[157] GAO, Pipeline Security and Safety: Improved Workforce Planning and Communication Needed, GAO-02-785, August 2002, p. 22.

[158] U.S. Department of Transportation, Office of Inspector General, May 21, 2008, p. 6.

[159] American Gas Association (AGA), American Petroleum Institute (API), Association of Oil Pipelines (AOPL), and American Public Gas Association (APGA), joint letter to members of the Senate Commerce Committee providing views on S. 1052, August 22, 2005.

[160] T.J. Barrett, Administrator, Pipeline and Hazardous Materials Safety Administration, Testimony before the Senate Committee on Commerce, Science, and Transportation on Federal Efforts for Rail and Surface Transportation Security, January 18, 2007.

[161] Sammon, John, Transportation Security Administration, Testimony before the House Transportation and Infrastructure Committee, Railroad, Pipelines, and Hazardous Materials Subcommittee hearing on Implementation of the Pipeline Inspection, Protection, Enforcement, and Safety Act of 2006, June 24, 2008.

[162] TSA, Pipeline Security Division, personal communication, February 17, 2009.

[163] Jack Fox, General Manager, Pipeline Security Division, Transportation Security Administration (TSA), remarks before the Louisiana Gas Association Pipeline Safety Conference, New Orleans, LA, July 25, 2012.

[164] National Energy Board of Canada, Proposed Regulatory Change (PRC) 2010-01, Adoption of CSA Z246.1-09 Security Management for Petroleum and Natural Gas Industry Systems, File Ad-GA-SEC-SecGen 0901, May 3, 2010, p. 1, https://www.neb-one.gc.ca/ll-eng/livelink.exe/fetch/2000/90463/409054/614444/A1S7H7_-_Proposed_Regulatory_Change_(PRC)_2010-01.pdf?nodeid=614556&vernum=0.

[165] National Transportation Safety Board, NTSB Issues Urgent Safety Recommendations as a Result of Preliminary Findings in San Bruno Pipeline Rupture Investigation; Hearing Scheduled For March," SB-11-01, press release, January 3, 2011.

[166] Ibid.

[167] Linda Daugherty, March 2, 2011.

[168] Dina Cappielo, "What Lies Beneath," Houston Chronicle, November 12, 2006, p. A1.

[169] "Pig" is the common acronym for "pipeline inspection gauge."

[170] Pete Carey, "Pipeline Inspection Not an Exact Science," San Jose Mercury News, October 11, 2010.

[171] The Pipeline Safety Improvement Act of 2002 (P.L. 107-355) directed the DOT to issue regulations on using internal inspection, pressure testing, and direct assessment to natural gas pipelines in high consequence areas.

[172] Christina Sames, Vice President, American Gas Association, November 4, 2010.

[173] John Kiefner, "Overview of Hydrostatic Testing," presentation to the Hydrostatic Testing Symposium, California Public Utilities Commission, May 6, 2011, http://www.cpuc.ca.gov/NR/rdonlyres/1A47C67C-4398-49CA-B52A-A8B5CD13457B/0/HydrostaticTestingSymposiumPresentationMaterialsversiontopost.pdf.

[174] For an example of such a review, see "The Northern Great Plains at Risk: Oil Spill Planning in Keystone Pipeline System, Plains Justice," Billings, MT, November 23, 2010.

[175] Sharon Theimer, "Government Lacks Copies of Emergency Response Plans Developed by Gas Pipeline Operators," Associated Press, October 6, 2010.

[176] Andrew Black, President, Association of Oil Pipe Lines, Remarks at the Different Pathways to a Common Goal: PIPA, Damage Prevention, & Greater Public Awareness and Involvement Conference, Pipeline Safety Trust, New Orleans, LA, November 5, 2010.

[177] Rob Rogers, "Photo Shows Broken Silvertip Pipeline in Yellowstone River," Billings Gazette, September 7, 2011.

[178] Cynthia L. Quarterman, Administrator, Pipeline and Hazardous Materials Safety Administration, testimony before the House Committee on Transportation and Infrastructure, Subcommittee on Railroads, Pipelines, and Hazardous Materials Hearing on the Silvertip Pipeline Oil Spill in Yellowstone County, Montana," July 14, 2001, p. 4.

[179] American Gas Association, "Utilities Work Towards Increased Safety for Natural Gas Infrastructure," press release, November 14, 2012.

[180] Interstate Natural Gas Association of America, "Interstate Natural Gas Pipelines Committed to Safety," press release, November 14, 2012.

[181] Association of Oil Pipelines, "Pipeline Operators Welcome NTSB Focus on Pipeline Safety," press release, November 14, 2012.

INDEX

D

E

F

N

O

P

Q

R

S

W

Y